Fuzzy Management Methods

Series editors

Andreas Meier, Fribourg, Switzerland
Witold Pedrycz, Edmonton, Canada
Edy Portmann, Bern, Switzerland

More information about this series at http://www.springer.com/series/11223

Nicolas Werro

Fuzzy Classification
of Online Customers

 Springer

Nicolas Werro
Villars-sur-Glane
Switzerland

ISSN 2196-4130 ISSN 2196-4149 (electronic)
Fuzzy Management Methods
ISBN 978-3-319-15969-0 ISBN 978-3-319-15970-6 (eBook)
DOI 10.1007/978-3-319-15970-6

Library of Congress Control Number: 2015932422

Springer Cham Heidelberg New York Dordrecht London

Printed on acid-free paper

Springer International Publishing AG Switzerland is part of Springer Science+Business Media
(www.springer.com)

To my love Delia and my parents Monique and René for their patience and support during these last years.

Foreword

Over the past decade, with the incredible rise of the Internet in general and e-commerce in particular, new and interesting avenues of research have opened in the broad field of Customer Relationship Management. The problem goes beyond the establishment of a sophisticated relational database coupled with a user-friendly web interface for managing the customers, their orders and the company stocks. Indeed, thanks to the large amount of personal data collected on customers, particularly those using the web for their purchases, it becomes possible to highly customize their management in order to retain them. To achieve this objective, it is necessary to segment (classify) customers based on various criteria in order to offer them benefits such as discounts, gift certificates or other promotions. In this context, traditional (sharp 0, 1) classification techniques lack of nuance and often can only register the damage when a loyal customer goes "brutally" in the lost ones category. To overcome this problem, the author proposes to combine relational databases practices commonly used for CRM with fuzzy logic techniques. Thus, it becomes possible for a client to belong to the category of loyal customers with a given "context", the evolution of which could be monitored in order to intervene early enough.

Convinced by the basic idea, that it is possible to improve online businesses CRM by using a classification based on fuzzy logic, the author adopts a four-step approach, fairly typical of applied research:

- First, in Chaps. 2 and 3, he consolidates its basic idea by rooting it in a proven theoretical framework. The relational model is extended with the notion of contexts and a few new operators. Thus, by using linguistic variables, it becomes possible to extend the classical schema of a relational database with notions of fuzzy clustering proper to a given domain. fCQL, a query language database for the formulation of fuzzy queries, (originally proposed by Schindler and slightly extended by the author), is also presented and briefly compared to others.
- Secondly, he specifies the application domain of his novel idea by introducing the reader to CRM (Chap. 4) and by presenting several examples illustrating the advantages of a fuzzy classification over sharp (0, 1) one.

- In Chap. 6, he introduces a quite convincing and complete real case study of a fuzzy customers classification within the domain of online shops.
- Finally, he presents in Chap. 7 the ambitious fCQL software toolkit, which has the goal of supporting and automatizing his new fuzzy approach of classification, while remaining compatible with classical database legacy systems.

To conclude, I want to strongly state that Nicolas Werro's book represents a very interesting applied research contribution. The concepts are original and well introduced; the examples and the case study are quite convincing; and the fCQL software toolkit is an excellent and very well thought engineered piece of software.

Prof. Dr. Jacques Pasquier-Rocha
Chair of Software Engineering
Department of Informatics
University of Fribourg-CH

Acknowledgments

My first thought goes to Prof. Dr. Andreas Meier, my supervisor, who gave me the idea and the opportunity of writing this thesis in the fascinating fields of fuzzy logic, database systems, CRM and E-Business. His continuous support and his encouragements have made the achievement of this thesis possible. I owe him the publication of many common research papers as well as five beautiful years of teamwork.

I am very grateful to Prof. Dr. Jacques Pasquier, my second supervisor, for his support, his careful reading and his valuable feedback. As supervisor of my Diploma thesis, he is undoubtedly one of the reasons for me to write a Ph.D. Thesis.

Warm thanks go to my colleagues, Dr. Henrik Stormer, Christian Mezger, Daniel Frauchiger, Darius Zumstein, Marco Savini, Daniel Risch, Dr. Günter Schindler, Martin Albrecht and Miltiadis Sarakinos, who actively participated in my research and with whom I had the chance to coauthor several publications. I am also very thankful to my former students, Christian Nançoz, Anne Gachet and Christian Savary whose Master/Bachelor/Diploma thesis made an important contribution to this work.

I would like to express my gratitude to all my colleagues of the Information Systems Research Group as well as the staff of the Department of Informatics (DIUF) for their friendship and support. Writing a Ph.D. thesis is a long and lonely way which I would not have faced without their presence.

Last but not least, special thanks go to Dr. Dac Hoa Nguyen who carefully proofread this document. His numerous hints and corrections have been precious to improve the language as well as the clarity of this work.

Contents

Chapter 1
Introduction

This introductory chapter first provides the reader a motivation of the pertinence of the fuzzy classification approach in the fields of information systems, customer relationship management and e-business (see Sect. 1.1). Then, Sect. 1.2 enumerates the research objectives which are treated in this thesis. Finally Sects. 1.3 and 1.4 give an outline of the thesis and mention the published contributions which are part of this work.

1.1 Motivation

In the last decades, information systems have revolutionized the way information can be stored and processed. As a result, the information volume has significantly increased leading to an information overload. It therefore becomes difficult to analyze the large amounts of available data and to generate appropriate management decisions. In practice, information systems mostly use relational databases in order to store these data collections. Another issue, using the relational model, is the restriction of having sharp, precise data and therefore a dichotomous querying process which is not well suited for decision making.

In this context, this thesis proposes a fuzzy classification approach which combines relational databases with fuzzy logic. This approach reduces the complexity of the data by classifying similar elements together and, at the same time, extracts additional valuable information by having fuzzy classes. The proposed fuzzy classification is achieved by an extension of the relational database schema, therefore it can be used with every relational database product and does not require any transformation or migration of the business data. The querying process of a fuzzy classification is also much more appropriate for decision making in the sense that a user can formulate unsharp queries on a linguistic level. The introduced linguistic variables and terms encapsulate the complexity of the domain as well as the business definitions (e.g. profitable customer). This leads to an intuitive and human-oriented querying process incorporating the business requirements perspectives.

© Springer International Publishing Switzerland 2015
N. Werro, *Fuzzy Classification of Online Customers*, Fuzzy Management Methods,
DOI 10.1007/978-3-319-15970-6_1

A pertinent and promising application field of the proposed fuzzy classification approach is the customer relationship management (CRM). Since CRM puts the customers in the focus of the company's strategy, effective means for the evaluation of the customer relationships have to be used. The strategic objectives of CRM, which are the acquisition, the retention and the recovery of customers, can be better achieved using fuzzy customer classes rather than traditional (sharp) evaluation methods. For instance, using a fuzzy classification, companies can derive the precise value of their customers as well as semantic information revealing their potential and their possible weaknesses. This additional information is pertinent for the acquisition (e.g. marketing campaigns), the development (e.g. cross/up-selling via personalized offers), the retention (e.g. personalized privileges) and the recovery of customers (e.g. churning customer detection). Furthermore, the controlling loop can be greatly improved by using a hierarchical fuzzy classification.

Building and maintaining profitable customer relationships are important issues in the field of electronic commerce since the World Wide Web enables a global market in which customers can easily compare competitive offers. On the other hand, online customers have the advantages that many information about their behavior can be derived from online shop systems. Therefore the analysis of online customers is a good case study to assess the feasibility and the validity of the fuzzy classification approach.

1.2 Research Issues

This thesis has been realized following a design science approach, it therefore aims at first creating innovative concepts which improve the actual human and organizational capabilities, secondly, at evaluating these concepts by providing concrete instantiations (i.e. implemented or prototype systems) [50, 74]. According to this research paradigm, the objectives of this thesis are the following:

- The first objective of this thesis is to extend the querying ability of the fuzzy classification approach proposed by Schindler [102]. By adding new clauses to the fuzzy Classification Query Language, the user should be given more powerful means for selecting elements within a fuzzy classification (see Chap. 3).
- The second objective of this thesis is to study a pertinent and promising application domain in which the benefits and the integration possibilities of the fuzzy classification approach are clearly identified (see Chap. 4).
- The third objective is, considering the application domain specificities, to extend the original fuzzy classification approach by new concepts which provide additional capabilities to the system (see Chap. 5).
- To validate the proposed concepts, the fourth objective is to realize a case study where the extended fuzzy classification approach is implemented with real life conditions and data (see Chap. 6).

- The fifth and last objective is to provide a prototype, i.e. the fCQL toolkit, implementing the extended fuzzy classification approach proving this way that the proposed concepts are not only valid in real conditions but are also realizable, i.e. the benefits can really be achieved in the application domain (see Chap. 7).

1.3 Chapters' Overview

This thesis is organized in three main parts, each part containing two chapters. The first part relates to the conceptual perspective of this work, i.e. it contains the theoretical background on which the rest of the thesis is built, namely:

- *Chapter 2—Fuzzy Set Theory*: In this chapter, the main concepts and mathematical notions of the fuzzy set theory are exposed. It emphasizes the affinity between human beings and the fuzziness, and depicts how fuzzy sets can capture and model the subjectivity and/or uncertainty which are common in the natural language. It finally exposes several application fields in which the fuzzy set theory has been successfully applied, i.e. the possibility theory, the fuzzy control theory, the fuzzy expert systems, the fuzzy classification and the fuzzy database systems.
- *Chapter 3—Relational Databases & Fuzzy Classification*: This chapter introduces the fuzzy classification approach with relational databases and the associated query language fCQL. It first motivates the pertinence of achieving fuzzy classifications on top of relational databases, then it describes the necessary extensions of the relational model. It afterwards illustrates the fuzzy classification query language fCQL allowing the formulation of intuitive fuzzy queries by the use of linguistic variables and terms. Finally, it discusses other fuzzy approaches combining fuzziness and databases.

The second part of this work, the customer perspective, presents the application field in which the fuzzy classification approach is evaluated, i.e. the customer relationship management, and the new perspectives offered by fuzzy customer classes:

- *Chapter 4—Customer Relationship Management*: This chapter gives the reader an introduction of the customer relationship management. As CRM is a very broad topic, the chapter focuses on two aspects: CRM theoretical constructs like the customer value, lifetime value and equity as well as the customer satisfaction, loyalty and retention which are required in the next chapters, and the architecture of CRM systems in which the fuzzy classification approach can be integrated.
- *Chapter 5—Fuzzy Customer Classes*: In this chapter, the benefits of the fuzzy classes in the customer relationship management are developed. The chapter first looks at the precise customer positioning ability of fuzzy classes. This ability allows a better planning and optimization of marketing campaigns as well as an individual customer monitoring, for instance to detect churning customers. Then, personalization issues are treated since fuzzy classes can automate the mass customization. This opens the door to personalized values like an individual discount but also to personalized products and services by introducing fuzzy product

portfolios. This chapter finally treats the customer assessment and controlling which can be realized with the help of hierarchical fuzzy classifications.

The third and last part of this thesis, the application and implementation perspective, aims at proving the applicability of the previously proposed concepts in a real case study and at presenting a concrete implementation of the fuzzy classification approach:

- *Chapter 6—Fuzzy Classification applied to Online Shops*: This chapter demonstrates a concrete implementation of a hierarchical fuzzy classification of online customers. Since the case study takes place in the e-business field, this chapter shortly presents the e-business framework, the thematic of e-commerce and online shops, and then goes on the analysis of online customers. Based on the available information about online customers and real data from a small-sized wholesaler company, four customers are analyzed through the proposed hierarchy of fuzzy classifications.
- *Chapter 7—fCQL Toolkit*: In this chapter, an implementation of the fuzzy classification approach, the fCQL toolkit, is illustrated. The fCQL toolkit does not only cover the querying process of a (hierarchical) fuzzy classification but also encompasses the definition process and the results evaluation. The chapter first describes the architecture of the fCQL toolkit. Then it illustrates the fCQL toolkit's user interface through the successive stages which are the database connection, the preliminary data analysis, the fuzzy classification definition, the fuzzy querying process and the results evaluation. The chapter finally describes the structure of the meta-tables storing all the fuzzy classification definitions.

Finally, Chap. 8 summarizes the key aspects developed in this thesis and discusses further developments as well as other promising application domains.

1.4 Published Work

Many aspects of the present thesis have been published in international conferences, journals and in a handbook. Contributions concerning the fuzzy classification approach, its application to customer relationship management and the implementation aspects have been published in [80, 82–84, 114, 123, 125, 126].

Since the case study of this thesis has been realized in the e-business field with online customers, a related topic are online shops and more precisely online shops for SME's. In this field several contributions have been published in [37, 113, 124, 127].

Part I
Conceptual Perspective

Chapter 2
Fuzzy Set Theory

This chapter aims to present the main concepts and mathematical notions of the fuzzy set theory (also called fuzzy logic or fuzzy logic theory[1]) which are necessary for the understanding of this work. It also gives a non exhaustive overview of the application domains where the fuzziness has been successfully applied. For this purpose, Sect. 2.1 introduces the fuzzy logic with respect to the human beings then Sect. 2.2 describes the concept of fuzzy sets, Sects. 2.3 and 2.4 define the main fuzzy sets' properties and operations, finally Sect. 2.5 presents some of the main application areas of the fuzzy set theory.

2.1 Human Beings and Fuzziness

The fuzzy set theory has been proposed in 1965 by Lofti A. Zadeh from the University of Berkeley [131]. This theory is based on the intuitive reasoning by taking into account the human subjectivity and imprecision. It is not an imprecise theory but a rigorous mathematical theory which deals with subjectivity and/or uncertainty which are common in the natural language. The natural language is a very complicated structure which is fundamental, not only in the human communication, but also in the way human beings think and perceive the surrounding world [94]. The main idea of the fuzzy logic is to capture the vagueness of the human thinking and to express it with appropriate mathematical tools [44]. More precisely, "the fuzzy logic provides a mathematical power for the emulation of the higher order cognitive functions, the thought and perception" [46].

Unlike computers, the human reasoning is not binary where everything is either yes (true) or no (false) but deals with imprecise concepts like 'a tall man', 'a moderate temperature' or 'a large profit'. These concepts are ambiguous in the sense that they cannot be sharply defined. For instance, the question whether a person is tall cannot

---

[1] Although fuzzy logic is an application of the fuzzy set theory extending the boolean logic, it is often used as a generic term encompassing the fuzzy set theory and its applications.

© Springer International Publishing Switzerland 2015
N. Werro, *Fuzzy Classification of Online Customers*, Fuzzy Management Methods,
DOI 10.1007/978-3-319-15970-6_2

be universally answered as some people will agree and others won't. Despite the fact that the definition of the word 'tall' is clear, it is not possible to sharply state if a person is tall because the answer may depend on the individual perception. Even for one person it may not be possible to give a clear and precise answer as the belonging to a concept (e.g. tall person) is often not sharp but fuzzy, involving a partial matching expressed in the natural language by the expressions 'very', 'slightly', 'more or less', etc.

Example 2.1 In order to illustrate this ambiguity, consider the concept 'middle-aged' for a person. This concept is clear in the people's mind however it is difficult to explicitly determine the precise beginning and ending years for a middle-aged person. Once again every individual might give a different definition of that concept. Let's assume that a survey states that a middle-aged person is between 35 and 55 years old. The concept 'middle-aged' can then be represented as a set illustrated in Fig. 2.1. This set represents the truth function of the concept 'middle-aged' according to the survey where the X-axis represents the age and the Y-axis contains the truth value. A truth value of 1 means that the age corresponds to the concept and a value of 0 indicates the age does not belong to that concept. This definition of the concept 'middle-aged' implies that a person who is 34 years old would suddenly become middle-aged on his next birthday. Similarly, just after his 56th birthday this person would no longer be middle-aged. This definition of the concept 'middle-aged' seems therefore unnatural as it does not match the human perception due to the sharply fixed boundaries.

A way of better modeling the imprecision of the human thinking is to introduce the notion of partial membership which allows a continuous transition between the different concepts. The notion of partial belonging can be represented by a fuzzy set (see Fig. 2.2). Fuzzy sets are the foundation of the fuzzy logic theory and are presented in Sect. 2.2. With this new definition a person enters the concept 'middle-aged' at the age of 20 with a continuous increment till the full belonging at the age of 45 and then progressively quits the concept. This way, there are no more extreme steps (sharp boundaries) such that just within a year somebody jumps into or out of the concept 'middle-aged'.

Fig. 2.1 Sharply defined concept of a middle-aged person

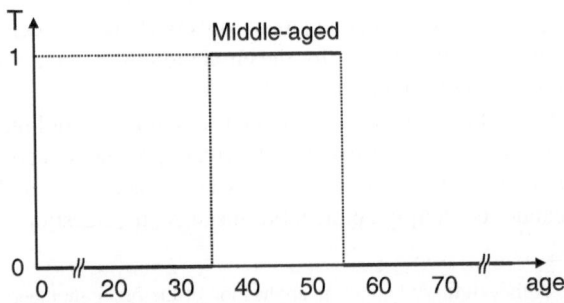

Fig. 2.2 Concept of a middle-aged person defined with a fuzzy set

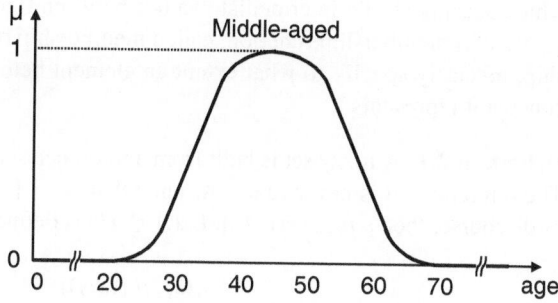

The ambiguity is part of the human thinking and is ubiquitous in the natural language. Different aspects of ambiguity can be distinguished [88]:

- *Incompleteness*: The ambiguity is caused by a lack of information or knowledge. For example, a sentence in a foreign language cannot be understood if the given language is unknown.
- *Homonymy*: A word with several possible meanings might be ambiguous if the correct interpretation is not clear. The word 'trailer' for instance has two very different definitions; it can be a motion picture preview or a vehicle depending on the context in which it is mentioned.
- *Randomness*: The ambiguity comes from the fact that the result of an event is not known since it will happen in the future. An example is when a dice is thrown and it is not yet known which side will show up. This aspect of ambiguity is covered in mathematics by the probability theory.
- *Imprecision*: An information can be ambiguous due to its imprecision, i.e. the information contains errors or noise and is not exact.
- *Fuzziness*: This aspect covers the ambiguity with respect to words, that is to say the ambiguity of semantics. For instance, it is ambiguous whether a person is tall.

The fuzzy logic theory deals with the last kind of ambiguity, the fuzziness [88]. It proposes mathematical notions to model the imprecision of the human thinking. Considering that the fuzziness is ubiquitous and essential for the human beings, the fuzzy logic theory offers new perspectives for improving the human-machine interactions. One important aspect of this thesis is the ability of processing intuitive and human-oriented queries based on linguistic terms or expressions.

2.2 Concept of Fuzzy Sets

The fuzzy logic theory is based on fuzzy sets which are a natural extension of the classical set theory. A sharp set (also called crisp set) is defined by a bivalent truth function which only accepts the values 0 and 1 meaning that an element fully belongs to a set or does not at all, whereas a fuzzy set is determined by a membership function which accepts all the intermediate values between 0 and 1 (see Example 2.1). The values of a membership function, called membership degrees or grades of membership, precisely specify to what extent an element belongs to a fuzzy set, i.e. to the concept it represents.

Definition 2.1 A fuzzy set is built from a reference set called *universe of discourse*. The reference set is never fuzzy. Assume that $U = \{x_1, x_2, \ldots, x_n\}$ is the universe of discourse, then a *fuzzy set A* in U ($A \subset U$) is defined as a set of ordered pairs

$$\{(x_i, \mu_A(x_i))\}$$

where $x_i \in U$, $\mu_A : U \to [0, 1]$ is the *membership function* of A and $\mu_A(x) \in [0, 1]$ is the *degree of membership* of x in A.

Example 2.2 Consider the universe of discourse $U = \{1, 2, 3, 4, 5, 6\}$. Then a fuzzy set A holding the concept 'large number' can be represented as

$$A = \{(1, 0), (2, 0), (3, 0.2), (4, 0.5), (5, 0.8), (6, 1)\}$$

With the considered universe, the numbers 1 and 2 are not 'large numbers', i.e. the membership degrees equal 0. Numbers 3–5 partially belong to the concept 'large number' with a membership degree of 0.2, 0.5 and 0.8. Finally number 6 is a large number with a full membership degree.

It is important to note that the definition of the membership degrees is subjective and context dependent, meaning that each person has his own perception of the concept 'large number' and that the interpretation is dependent on the universe of discourse and the context in which the fuzzy set is used. In Example 2.2 for instance, the membership degrees of the elements would be quite different if the universe of discourse contained numbers up to 100 or even 1000. In a similar manner, the concept 'large profit' would have a distinct signification for a small and a large enterprise.

Fuzzy sets are commonly represented by a membership function. Depending on the reference set, the membership functions are either discrete or continuous. Figure 2.3 shows the truth function of a sharp set in comparison to the membership functions of a discrete and a continuous fuzzy set.

Usually, several fuzzy sets are defined on the same reference set forming a fuzzy partition of the universe. A linguistic expression from the natural language can label the fuzzy sets in order to express their semantics. In the case of Example 2.1, the reference set can hold the concepts 'young', 'middle-aged' and 'old' at the same time allowing a continuous transition between them (see Fig. 2.4). This construct is essential in the fuzzy logic theory and is called a linguistic variable. A linguistic variable is a variable whose values are words or sentences instead of numerical values [134, 135, 136]. These values are called terms (also linguistic or verbal terms) and are represented by fuzzy sets.

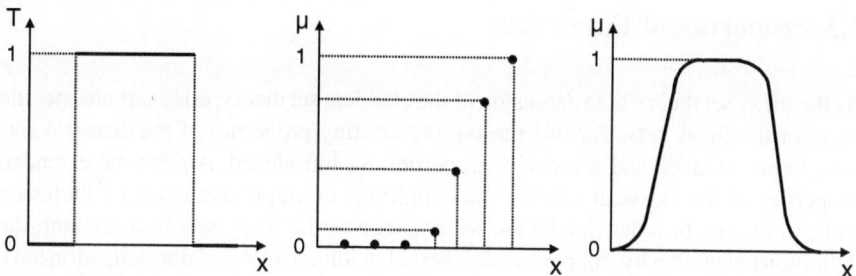

Fig. 2.3 Truth function of a sharp set and membership functions of a discrete and a continuous fuzzy set

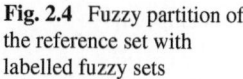

Fig. 2.4 Fuzzy partition of the reference set with labelled fuzzy sets

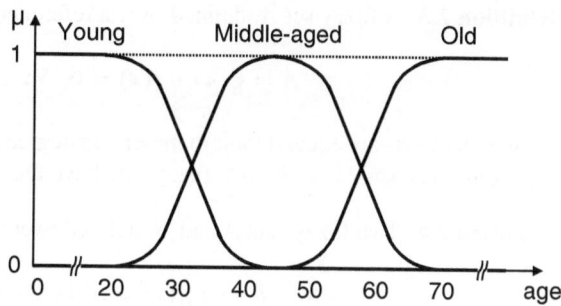

Definition 2.2 A *linguistic variable* is characterized by a quintuple

$$(X, T, U, G, M)$$

where X is the name of the variable, T is the set of terms of X, U is the universe of discourse, G is a syntactic rule for generating the name of the terms and M is a semantic rule for associating each term with its meaning, i.e. a fuzzy set defined on U [134, 135, 136].

Example 2.3 The linguistic variable represented in Fig. 2.4 is defined by the quintuple (X, T, U, G, M) where X is 'age', T is the set {young, middle-aged, old} generated by G and M specifies for each term a corresponding fuzzy set on the universe $U = [0, 100]$.

The ability of giving a partial belonging to the elements allows a continuous transition between the fuzzy sets instead of having sharply fixed boundaries. This way, it is possible to better reflect the reality where everything is not black or white but often differentiated by grey values. The definition of a fuzzy set can therefore adequately express the subjectivity and the imprecision of the human thinking. Furthermore, the concept of linguistic variable is the basis for representing the human knowledge within human oriented rules or queries which can be processed by computers.

2.3 Properties of Fuzzy Sets

As the fuzzy set theory is an extension of the classical set theory, crisp sets are specific cases of the fuzzy sets. For this reason, the existing properties of the classical sets have to be extended and some new properties are introduced. Among the extended properties of the classical sets are the definitions of emptiness, equality, inclusion and cardinality. In order to take the wider scope of the fuzzy sets into account, the definitions of convexity, support, α-cut, kernel, width, height and normalization have been introduced.

A fuzzy set is considered to be empty if the membership degrees of all the elements of the universe are equal to zero.

Definition 2.3 A fuzzy set A, defined over a reference set U, is *empty* if

$$A = \varnothing \Leftrightarrow \mu_A(x) = 0, \ \forall x \in U$$

Two fuzzy sets are equal if their membership degrees are equal for all the elements of the reference set, i.e. if the two fuzzy sets have the same membership function.

Definition 2.4 Two fuzzy sets A and B, defined over a reference set U, are *equal* if

$$A = B \Leftrightarrow \mu_A(x) = \mu_B(x), \ \forall x \in U$$

A fuzzy set A is included in a fuzzy set B if the degrees of membership of A are smaller or equal to the membership degrees of B for all the elements of the universe (see Fig. 2.5).

Definition 2.5 Let A and B be two fuzzy sets defined over a reference set U, A is *included* in B if

$$A \subseteq B \Leftrightarrow \mu_A(x) \leq \mu_B(x), \ \forall x \in U$$

The cardinality of a crisp set equals the number of elements it contains. In a fuzzy set the elements can have a partial belonging, therefore the cardinality is the sum of the membership degrees of the reference set elements. If the reference set is infinite,

Fig. 2.5 Inclusion of the fuzzy set A in the fuzzy set B

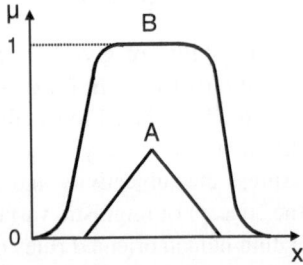

an integral over the universe is used instead of the addition. It is also possible to derive the relative cardinality of a fuzzy set by dividing the cardinality of the fuzzy set by the cardinality of the universe. The relative cardinality allows fuzzy sets to be compared if they are based on the same universe.

Definition 2.6 The *cardinality* and the *relative cardinality* of a fuzzy set A, defined over a finite universe U, are defined as

$$Card(A) = |A| = \sum_{x \in U} \mu_A(x)$$

$$RelCard(A) = ||A|| = \frac{|A|}{|U|}$$

Example 2.4 Consider the fuzzy set A of Example 2.2.

$$A = \{(1, 0), (2, 0), (3, 0.2), (4, 0.5), (5, 0.8), (6, 1)\}$$

Then the cardinality and the relative cardinality of A are

$$Card(A) = 0 + 0 + 0.2 + 0.5 + 0.8 + 1 = 2.5$$

$$RelCard(A) = \frac{2.5}{6} \approx 0.417$$

Generally, linguistic notions are represented by a convex fuzzy set (see Fig. 2.6). A fuzzy set is convex if any point located between two other points has a higher membership degree than the minimum membership degree of these points.

Definition 2.7 A fuzzy set A defined over a reference set U is *convex* if

$$\forall x, y \in U, \forall \lambda \in [0, 1] : \mu_A(\lambda x + (1 - \lambda)y) \geq min(\mu_A(x), \mu_A(y))$$

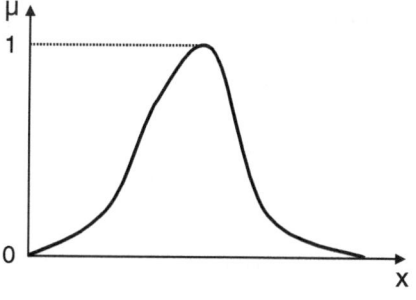 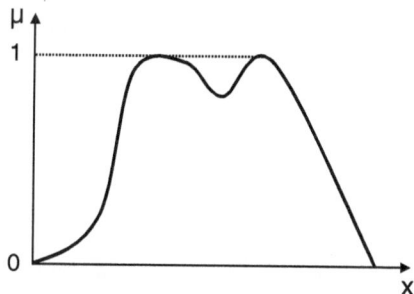

Fig. 2.6 Convex and non-convex fuzzy sets

Fig. 2.7 Support, α-cut and kernel of a fuzzy set

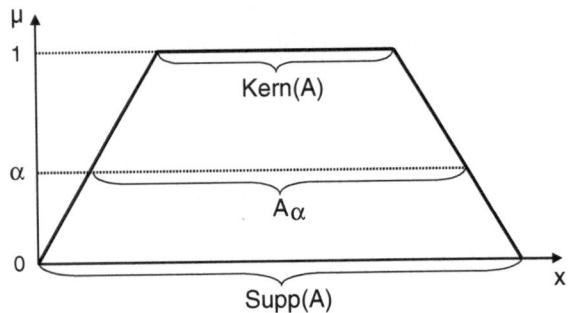

The support of a fuzzy set is the sharp subset of the universe where the membership degrees are greater than zero (see Fig. 2.7).

Definition 2.8 The *support* of a fuzzy set A defined over a reference set U is a crisp subset of U that complies with

$$Supp(A) = \{x \in U, \ \mu_A(x) > 0\}$$

The α-cut, resp. the strong α-cut, of a fuzzy set is the crisp subset of the universe where the membership degrees are greater or equal, resp. greater, than the specified α value (see Fig. 2.7).

Definition 2.9 The α-*cut* and the *strong* α-*cut* of a fuzzy set A, defined over a reference set U, are a crisp subset of U that complies with

$$A_\alpha = \{x \in U, \ \mu_A(x) \geq \alpha, \ \alpha \in [0, 1]\}$$

$$A'_\alpha = \{x \in U, \ \mu_A(x) > \alpha, \ \alpha \in [0, 1]\}$$

The kernel of a fuzzy set is the crisp subset of the universe where the membership degrees are equal to 1 (see Fig. 2.7).

Definition 2.10 The *kernel* of a fuzzy set A defined over a reference set U is a crisp subset of U that complies with

$$Kern(A) = \{x \in U, \ \mu_A(x) = 1\}$$

The width of a convex fuzzy set is the length of the support, which in the case of a convex fuzzy set is an interval.

Definition 2.11 The *width* of a convex fuzzy set A with support $Supp(A)$, defined on a bounded reference set, is defined as

$$Width(A) = max(Supp(A)) - min(Supp(A))$$

The height of a fuzzy set is the maximum membership degree of all the elements of the universe.

Definition 2.12 The *height* of a fuzzy set A defined on a bounded reference set U is defined as

$$Hgt(A) = max_{x \in U}(\mu_A(x))$$

A fuzzy set is said to be normalized if at least one element of the universe has a membership degree equal to 1.

Definition 2.13 A fuzzy set A defined over a reference set U is *normalized* if and only if

$$\exists x \in U, \ \mu_A(x) = Hgt(A) = 1$$

2.4 Operations on Fuzzy Sets

The operations of complement, intersection and union of the classical set theory can also be generalized for the fuzzy sets. For these operations, several definitions with different implications exist. This section only presents the most common operators from the Zadeh's original proposition [131]. Further operators can be found in Appendix A.

The complement of a fuzzy set is 1 minus the membership degrees of the elements of the universe. This definition respects the notion of strong negation [41].

Definition 2.14 The *complement* of a fuzzy set A defined over a reference set U is defined as

$$\neg A = \mu_{\neg A}(x) = 1 - \mu_A(x), \ x \in U$$

For the intersection (resp. the union), Zadeh proposes to use the minimum operator (resp. the maximum operator). These operators have the advantages of being easily understandable and very fast to compute. The intersection (resp. the union) of two fuzzy sets is the minimum (resp. the maximum) value of the membership degrees of the two fuzzy sets for all the elements of the reference set (see Fig. 2.8).

Definition 2.15 The *intersection* of two fuzzy sets A and B defined over a reference set U is defined as

$$A \cap B = \mu_{A \cap B}(x) = \mu_A(x) \wedge \mu_B(x) = min(\mu_A(x), \mu_B(x)), \ x \in U$$

Definition 2.16 The *union* of two fuzzy sets A and B defined over a reference set U is defined as

$$A \cup B = \mu_{A \cup B}(x) = \mu_A(x) \vee \mu_B(x) = max(\mu_A(x), \mu_B(x)), \ x \in U$$

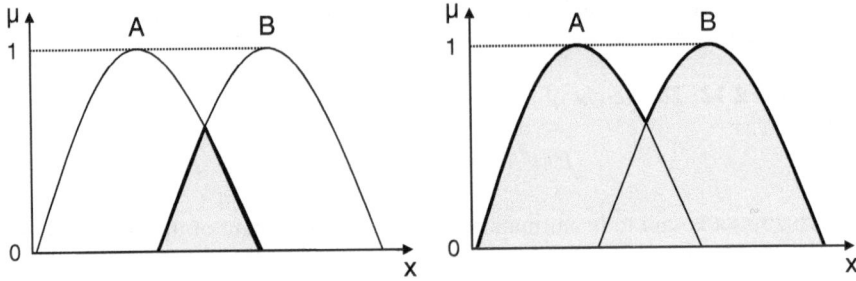

Fig. 2.8 Intersection and union of two fuzzy sets

Based on the intersection definition of two fuzzy sets, it is possible to introduce the notion of possibility (also called consistency or consensus) which is the fundament of the possibility theory which is briefly treated in Sect. 2.5.1. The possibility of two fuzzy sets, which determines the agreement degree between the concepts represented by the fuzzy sets, measures to what extent the fuzzy sets superpose each other and is defined as the highest membership degree of the intersection of these two fuzzy sets [137]. In the case of Zadeh's definition of the intersection this operation is called the max-min operation since it considers the maximum of the minimum values of the fuzzy sets.

Definition 2.17 The *possibility* of two fuzzy sets A and B defined over a reference set U is defined as

$$Poss(A, B) = max_{x \in U}(\mu_{A \cap B}(x)) = max_{x \in U}(min(\mu_A(x), \mu_B(x)))$$

Example 2.5 Consider the universe U and the fuzzy sets A and B.

$$U = \{a, b, c, d, e\}$$

$$A = \{(a, 0.4), (b, 0.8), (c, 1.0), (d, 0.8), (e, 0.2)\}$$

$$B = \{(a, 0.0), (b, 0.5), (c, 0.3), (d, 0.9), (e, 1.0)\}$$

Then the complement of A, the intersection, the union and the possibility of A and B are

$$\neg A = \{(a, 0.6), (b, 0.2), (c, 0.0), (d, 0.2), (e, 0.8)\}$$

$$A \cap B = \{(a, 0.0), (b, 0.5), (c, 0.3), (d, 0.8), (e, 0.2)\}$$

$$A \cup B = \{(a, 0.4), (b, 0.8), (c, 1.0), (d, 0.9), (e, 1.0)\}$$

$$Poss(A, B) = 0.8$$

It has to be noted that in contrast to the classical set theory, the intersection (resp. the union) of a fuzzy set and its complement does not result in the empty set (resp. in the universe).

$$A \cap \neg A = \{(a, 0.4), (b, 0.2), (c, 0.0), (d, 0.2), (e, 0.2)\} \neq \varnothing$$

$$A \cup \neg A = \{(a, 0.6), (b, 0.8), (c, 1.0), (d, 0.8), (e, 0.8)\} \neq U$$

More generally, the family of operators implementing the intersection is called *Triangular Norm* (abbreviated t-norm) and *Triangular Conorm* (abbreviated t-conorm or s-norm) for the union. These families comply with the properties showed in Table 2.1 [92].

As noted in Example 2.5, the t-norm and t-conorm operators do not comply with the Aristotle's laws of non-contradiction[2] and excluded middle[3] in order to express the vagueness of the human thinking [44]. The t-norm and t-conorm families are related by a general relation expressed in Definition 2.18 [1].

Definition 2.18 Let A and B be two fuzzy sets over the universe U, t a t-norm operator and s a t-conorm operator, then t and s are connected by the relation:

$$\mu_A(x) \, t \, \mu_B(x) = 1 - ((1 - \mu_A(x)) \, s \, (1 - \mu_B(x))), \ x \in U$$

The t-norms and t-conorms are also known as *non compensatory* operators meaning that there is no compensation effect between the elements. For instance, the results of a t-norm (resp. t-conorm) operator has an upper (resp. lower) limit defined by the minimum (resp. maximum) operator (see Fig. 2.9). The notion of compensation has an important significance for human beings who instinctively weigh up elements, especially in the context of decision making. For this reason, there exist so-called *averaging* and *compensatory* operators which do not comply with all the properties of t-norms and t-conorms [29].

Table 2.1 Properties of t-norm and t-conorm operators

Property	T-norm	T-conorm
Identity	$1 \wedge x = x$	$0 \vee x = x$
Commutativity	$x \wedge y = y \wedge x$	$x \vee y = y \vee x$
Associativity	$x \wedge (y \wedge z) = (x \wedge y) \wedge z$	$x \vee (y \vee z) = (x \vee y) \vee z$
Monotonicity	if $v \leq w$ and $x \leq y$ then $v \wedge x \leq w \wedge y$	$v \vee x \leq w \vee y$

---

[2] The law of non-contradiction states that the same thing cannot at the same time belong and not belong to the same object and in the same respect [2].

[3] The law of excluded middle states that of any subject, one thing must be either asserted or denied [2].

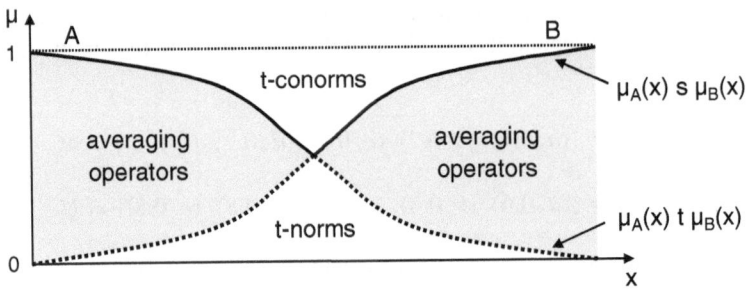

Fig. 2.9 t-norms, t-conorms and averaging operators

Averaging operators for intersection (resp. union) are more optimistic (resp. pessimistic) than t-norms (resp. t-conorms); their results are bounded by the chosen t-norm t and t-conorm s operators as shown in Fig. 2.9. In other words, "averaging operators realize the idea of trade-offs between conflicting goals when compensation is allowed" [139]. These operators have been empirically tested and proved to be well suited for modeling aggregation in a human decision environment [118].

Examples of averaging operators are the 'fuzzy and' and 'fuzzy or' proposed by Werners [122]. They are a combination of the minimum or maximum operator and the arithmetical mean weighted by a γ-argument. For $\gamma = 1$, the results equal the minimum, respectively the maximum, operator. For $\gamma = 0$, the results equal the arithmetical mean for both the 'fuzzy and' and the 'fuzzy or'.

Definition 2.19 The *fuzzy and* of two fuzzy sets A and B defined over a reference set U is defined as

$$\mu_{A \cap B}(x) = \gamma \, min(\mu_A(x), \mu_B(x)) + \frac{(1 - \gamma)}{2} (\mu_A(x) + \mu_B(x))$$

where $\gamma \in [0, 1]$ and $x \in U$.

Definition 2.20 The *fuzzy or* of two fuzzy sets A and B defined over a universe U is defined as

$$\mu_{A \cup B}(x) = \gamma \, max(\mu_A(x), \mu_B(x)) + \frac{(1 - \gamma)}{2} (\mu_A(x) + \mu_B(x))$$

where $\gamma \in [0, 1]$ and $x \in U$.

More interesting are the compensatory operators which have a compensation mechanism to reflect the human reasoning. Compensatory operators are located somewhere in between the intersection and union operators. An important compensatory operator is the γ-operator (also called 'compensatory and') which has been suggested as 'compensatory' and empirically tested by Zimmermann and Zysno [140]. It is composed by the algebraic product operator, a t-norm, and its counterpart the algebraic sum, a t-conorm following Definition 2.18 (see Appendix A). This operator has a γ-argument ranging from 0 to 1 which specifies whether the results should go in the direction of the algebraic product (t-norm) or the algebraic sum (t-conorm). The γ-argument therefore determines the strength the compensation mechanism.

Definition 2.21 The γ-*operator* of m fuzzy sets $A_1, \ldots, A_m$ defined over a reference set U with membership functions $\mu_1, \ldots, \mu_m$ is defined as

$$\mu_{A_i,comp}(x) = \left(\prod_{i=1}^{m} \mu_i(x) \right)^{(1-\gamma)} \left(1 - \prod_{i=1}^{m} (1 - \mu_i(x)) \right)^{\gamma}, \ \gamma \in [0, 1] \text{ and } x \in U$$

Another interesting kind of operators are the *linguistic modifiers* (also called linguistic hedges) which expresses linguistic notions like 'a little', 'slightly', 'very', 'extremely', 'more or less', etc. [132]. These operators are called fuzzy sets modifiers as they slightly modify the shape of a membership function according to the expressed notion. For instance, if a fuzzy set expresses the notion of 'young', a new fuzzy set with semantics 'very young' can be created by applying the linguistic modifier 'very' (see Fig. 2.10).

Once again, several operators can be used for the different linguistic modifiers. The most common operator for the modifier 'very' is the concentration which is the square value of the original membership degrees. Other linguistic modifiers can be found in Appendix A.

Definition 2.22 The *concentration* of a fuzzy set A defined on a universe U is defined as

$$\mu_{CON(A)}(x) = \mu_A^2(x), \ x \in U$$

There exist many other concepts based on fuzzy sets which cannot be presented here, namely the notion of distance between fuzzy sets [96], the notion of necessity and compatibility measures [137], the notion of fuzzy numbers, relations and similarities [131], the extension principle [134, 135, 136], etc. A general introduction to these concepts can be found in [41, 88, 139]. Note also that a detailed discussion of the properties of the operators listed in Appendix A can be found in [31, 66, 68, 122, 129, 139, 140].

Fig. 2.10 Fuzzy set with the linguistic modifier 'very'

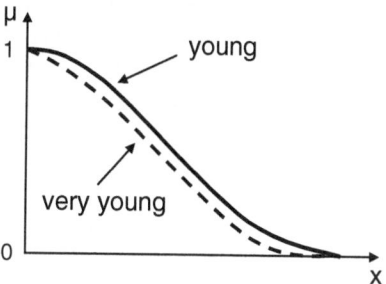

2.5 Application Fields

The fuzzy set theory has been successfully applied in various domains. The most important application areas are the fuzzy control, the fuzzy diagnosis, the fuzzy data analysis and the fuzzy classification [85]. This section aims to explicit the implication of the fuzzy set theory in some of these domains. First, Sect. 2.5.1 introduces the possibility theory as a basis for the approximate reasoning which allows the integration of the natural language into the reasoning process. Based on the approximate reasoning, Sect. 2.5.2 presents the fuzzy control theory in comparison to the modern control theory. Examples of fuzzy diagnosis and fuzzy data analysis areas are fuzzy expert systems depicted in Sect. 2.5.3 and the fuzzy classification approach presented in Sect. 2.5.4. Last but not least, Sect. 2.5.5 presents different approaches which enable the representation and the storage of the imprecision, i.e. fuzzy databases systems.

Nowadays a large number of real-world applications take advantages of the approximate reasoning [26]. Many other applications fields could have been discussed like the neural networks, the genetic algorithms, the evolutionary programming, the chaos theory, etc., but their presentation is beyond the scope of this thesis.

2.5.1 Possibility Theory

A membership value of a fuzzy set has been defined as the degree to which an element belongs to this fuzzy set. It is possible to give other interpretations to the membership degree like a certainty factor, a degree of truth, a degree of satisfaction and a degree of possibility [88]. In 1978 Zadeh extended the fuzzy set theory to a possibility theory where the membership values are considered as degrees of possibility. Zadeh justifies the possibility theory by the fact that "the imprecision that is intrinsic in natural languages is, in the main, possibilistic rather than probabilistic in nature" [137]. In contrast to the statistical perspective of the information which is involved in the coding, the transmission and the reception of the data [110], the theory of possibility focuses on the meaning of the information.

One central concept in the possibility theory is the possibility distribution which is the counterpart of the probability distribution in the probability theory. A possibility distribution is a fuzzy set called fuzzy restriction, which acts as an elastic constraint, whose membership function determines the compatibility or the possibility with the concept of the fuzzy set. Given a possibility distribution it is possible to compute the possibility of another fuzzy set defined on the same universe (see Definition 2.17). Consider for instance the possibility distribution 'young' of a linguistic variable 'age' defined on the universe U and the fuzzy set 'around 35' also defined on U. By knowing that 'Mary is young' it is then possible to calculate the possibility that 'Mary is around 35'. Note that the possibility represents a degree of feasibility whereas the probability is related to a degree of likelihood implying that what is possible might not be probable and, conversely, what is improbable might not be impossible [69].

The possibility theory opens the door to the fuzzy reasoning which can represent and manipulate the natural language. Almost all human related problems are so complex and so vague that only approximate linguistic expression can be used [69]. The fuzzy approximate reasoning is based on different fuzzy inference patterns which deal with different implication interpretations and also determine the way the uncertainties are propagated. The fuzzy inference can then compute or deduct elastic constraints (fuzzy sets) determined by membership functions via the possibility concept.

2.5.2 Fuzzy Control Theory

One of the reasons the scientific community took an interest in the fuzzy logic theory is the financial success of fuzzy control in home appliances in the Japanese industry. In 1990, the consumer products market using fuzzy controllers was estimated to 2 billion dollars [43]. Interestingly enough L.A. Zadeh is a major contributor of the modern control theory. The control theory is a very precise and strict approach in order to model systems or phenomena. As all the aspects of the model have to be specified, modeling a complicated system is an extensive operation. For example, an application could be used to predict the path of a hurricane but if it has to be developed from scratch, the hurricane will be gone before the application is ready to use. In the control theory, the number of processes to be implemented grows exponentially relatively to the number of variables defining the system [88]. For this reason some systems cannot be modeled even by high speed computers. A solution to this problematic is to roughly define systems with the help of the fuzzy logic theory. The fuzzy control is based on the approximate reasoning which offers a more realistic framework for human reasoning than the two-valued logic [134, 135, 136]. The main advantages of fuzzy control over the classical control theory is its ability of implementing human expert knowledge, its methods for modeling non-linear systems and a shorter time to market development [29, 43].

In the control theory systems are characterized by input and output variables as well as a set of rules. These rules define the behavior of the system. The output variables are then calculated by inference based on the input variables and the given rules. An inference is the construct 'A implies B, B implies C then A implies C'. When a premise 'X implies Y' holds then Y is true if X is true and, conversely, X is false if Y is false. This is called a syllogism and a famous example is:

Implication: All men are mortal
Premise: Socrates is a man
Conclusion: Socrates is mortal

In the control theory the premises are defined by rules in the form 'If X is F Then Y is G' where X (resp. Y) is an input (resp. output) variable and F (resp. G) is a condition on X (resp. Y). Zadeh introduced in 1973 the compositional rule of inference [133] which extends the inference mechanism in order to take the fuzziness into account. In 1993, Fullér and Zimmermann demonstrated the stability property of the conclusion using the compositional rule of inference which states that a conclusion depends continuously on the premise when the t-norm defining the composition and the membership function of the premise are continuous [38]. This property guarantees that small changes in the membership function of the premise, eventually due to errors, can imply only a small deviation in the conclusion.

The fuzzy rules can then be expressed in the natural language by the use of linguistic variables [134, 135, 136]. Zadeh's fuzzy inference example where the conditions are expressed by the means of words is:

Implication: If a tomato is red then it is ripe
Premise: This tomato is very red
Conclusion: This tomato is very ripe

These words allow the fuzzy rules to integrate the semantics of the human knowledge and can be represented as fuzzy sets. The evaluation process of the fuzzy inference also differs from the classical control theory in the sense that all the rules involving a given output variable are computed simultaneously and their results are then merged in order to derive the value of the output variable. This is a major advantage over the classical control theory as it implies a compensation mechanism between the involved rules. As a result, a much smaller set of rules is required to model a system as the intermediate values of the input variables are dynamically interpolated from the existing rules. It also implies an inherent fault tolerance; consider that a rule has been erroneously implemented or that a hardware defect returns wrong results, the value of the output variable can be compensated by other rules defining this variable.

Many concrete applications using fuzzy control can be found. The most famous one is the opening in 1988 of a subway system in Sendai City (Japan) using the fuzzy control to accelerate and brake the trains more smoothly than a human driver. Compared to conventional control, this new approach achieved significant improvements in the fields of safety, riding comfort, accuracy of stop gap, running time and energy consumption [130]. Other concrete applications can be found in domestic appliances like washing machines and vacuum cleaners, in visual systems like camera auto focus and photocopiers, in embedded car systems like anti-lock braking systems, transmission systems, cruise control and air conditioning, etc. A review of fuzzy control applications can be found in [29].

2.5.3 Fuzzy Expert Systems

Expert systems are a successful example from the broad field of artificial intelligence. Expert systems are knowledge-based systems which can derive decision or conclusion based on an extensive knowledge on a particular domain. More precisely, "an expert system is a program that can provide expertise for solving problems in a defined application area in the way the experts do" [65]. This knowledge is represented in a set of 'If-Then' rules. By applying inferences on the specified rules, expert systems are able to derive optimal decisions.

A major problematic, however, is to convert the experts' knowledge into a set of 'If-Then' rules which are exact given that the human representation of the knowledge cannot be sharply determined. This drawback can be overcome by introducing the fuzziness. This is done by allowing the definition of fuzzy rules, i.e. rules with words determined by a membership function, and by applying the previously defined fuzzy inference. Just like in the fuzzy control, the fuzzy inference allows a dynamic compensation between the different fuzzy rules which results in the definition of a smaller set of rules. Fuzzy expert systems are usually involved when processes cannot be described by exact algorithms or when these processes are difficult to model with conventional mathematical models [49].

Although the rules definition and the inference mechanism of fuzzy expert systems are similar to those in fuzzy control, fuzzy expert systems do not come under the category of fuzzy control [85]. Fuzzy control applications (often called fuzzy controllers) work in a closed loop schema where the output variables, which are derived from the input variables, directly act on the considered object. The rules are then executed in cycles in order to maintain a system. In the case of fuzzy expert systems and, more generally, for fuzzy diagnosis, fuzzy data analysis and fuzzy classification systems, the output information of a fuzzy system is dedicated to a human user or a monitoring device and hasn't any impact on the object itself.

Earl Cox [26] has implemented different fuzzy expert systems which have been successfully applied to the following domains: transportation, managed health care, financial services, insurance risk assessment, database information mining, company stability analysis, multi-resource and multi-project management, fraud detection, acquisition suitability studies, new product marketing and sales analysis. By comparing fuzzy expert systems with conventional expert systems Cox stated that "generally, the final models were less complex, smaller, and easier to build, implement, maintain, and extend than similar systems built using conventional symbolic expert systems" [26].

2.5.4 Fuzzy Classification

The fuzzy classification is a natural extension of the traditional classification, the same way that the fuzzy sets extend the classical sets (see Sect. 2.2). In a sharp classification, each object is assigned to exactly one class, meaning that the membership degree of the object is 1 in this class and 0 in all the others. The belonging of the objects in the classes is therefore mutually exclusive. In contrast, a fuzzy classification allows the objects to belong to several classes at the same time; furthermore, each object has membership degrees which express to what extent this object belongs to the different classes.

Definition 2.23 Let O be an object characterized by a t-dimensional feature vector $\underline{x}_O$ of a universe of discourse U. Often U is the space R^t. Let $C_1, \ldots, C_n$ be a set of classes which is given a priori or has to be discovered. A *fuzzy classification* calculates a membership vector $M = \{m_1, \ldots, m_n\}$ for the object O. The vector element $m_i \in [0, 1]$ is the degree of membership of O in the class C_i [85].

In many real applications, a dichotomous assignment of an object in one class is often not possible as no unique conclusion can be derived from the object features and/or the object features cannot be exactly observed [85]. This is particularly true for problems related to the human evaluation, intuition, perception and decision making where the problem structure is not dichotomous [139]. The definition of the classes can be determined by using the knowledge of experts of the domain or can be automatically found by the use of data mining techniques like cluster analysis [85].

The fuzzy classification approach can be used for instance for diagnosis and for decision making support. In the case of a diagnosis system for ill persons, the classification procedure can derive the illness based on the symptoms of the patient or find a suitable therapy considering the illness of the patient [85, 88]. In a decision making process, the classification (also called segmentation depending on the context) is used to derive management decisions based on several characteristics of the objects. A major issue in this field is the complexity of the data, i.e. the abundance of information. This complexity is a source of uncertainty due to the limited capability of human beings to observe and handle large amounts of data simultaneously [91]. As in the management field a large number of objects described by many features is usually considered, the classification approach, by grouping similar objects into classes, results in a complexity reduction which enables a better situation analysis [102]. Furthermore, the fuzzy classification, in contrast to the classical one, by allowing objects to belong to several classes at the same time, reduces the complexity of the data and also provides a much more precise information about the classified elements.

The fuzzy classification concept given in Definition 2.23 specifies that a membership vector for each object in the different classes has to be calculated. There exist many approaches to derive the membership vector of the classified objects. In this thesis, a context based approach which extends the relational database schema to a fuzzy relational database schema has been chosen. Chapter 3 presents this approach which has been originally proposed by Schindler [102].

2.5.5 *Fuzzy Database Systems & ER Models*

Many contributions in the fields of database systems which study the representation and the processing of imprecise information can be found in the literature. In this context, the imprecision can be given several interpretations [41]: an information is *uncertain* if it is incomplete or fuzzy, *unknown* if no information is available or *undefined* if it is inapplicable to the predefined domain. Note that this definition of imprecision in data management overlaps the general aspects of ambiguity discussed in Sect. 2.1.

Different approaches besides fuzzy logic can be used to model the imprecision. In the field of relational databases, Codd first introduced the notion of *NULL* values which allows the modeling of unknown information [23], then extended it to distinguish between missing values which may be applicable (i.e. unknown) called A-mark or inapplicable (i.e. undefined) called I-mark [24, 25]. In order to encompass the uncertainty of an information, Grant proposed an extension of the relational model to accept not only null values but also interval values as tuple components [45].

Another approach to model the uncertainty of information is the development of statistical and probabilistic databases. For instance, Wong proposed a framework allowing the extraction of information from imprecise databases by statistical means [128]. Barbará et al. extended the relational model by a probabilistic model which allows probabilities to be associated with attributes values. This probabilistic data model also includes missing probabilities and can therefore capture uncertainty in data values [3]. Similarly, Cavallo and Pittarelli proposed a theory of probabilistic databases which is a generalization of the relational model to accommodate both probabilistic and relational data. By assigning all tuples of a relation a probability distribution, it is possible to know the probability that a tuple belongs to the relation [18].

In order to better model the imprecision in databases, the use of the fuzzy logic theory can be considered. For this purpose, two different architectural concepts can be differentiated, depending at which level the fuzziness is applied [87]. A first category of systems, the fuzzy database systems, directly integrates the fuzziness at the data level. In contrast, a second category of systems still works with relational database systems and specifies an additional upper layer where fuzzy queries can be formulated on top of the RDBMS. In this subsection fuzzy database systems are considered since fuzzy query languages for relational databases are treated in Chap. 3.

Following Galindo et al. [41], the basic fuzzy relational database model is an extension of the relational model by adding to each tuple a grade in the interval [0, 1]. This extension is very similar to the one of the probabilistic databases previously mentioned, however both models cannot be compared. As discussed in Sect. 2.5.1, the probability focuses on the likelihood of events and has specific constraints, e.g. the probability distribution has to sum up to 1, which is not the case in fuzzy databases where the grades can be assigned several meanings. The tuples' grade and their associated meaning can then be used in the querying process. The grades may represent the membership degree of the tuples in the relation, the dependence strength between two attributes as well as the fulfillment of a condition or the importance degree of the tuples of a relation [41]. This basic fuzzy relational model suffers however from not being able to represent imprecise information at the attribute level since grades are assigned to the tuples directly [41].

In 1986 Zvieli and Chen [143] proposed an extension of the entity-relationship model (ER) [21] by allowing the definition of fuzzy attributes, entities and relationships. In this approach, three levels of fuzziness in the ER model can be distinguished:

- On a first level, entities, relationships and attributes include a membership degree to the model. This degree expresses the importance of the element to the model such that given an importance threshold, some elements can be eliminated.
- On a second level, instances of entities and relationships can be fuzzy. In this case, a degree (whose meaning has to be specified) depicts the belonging of a tuple (resp. a relationship instance) in the entity (resp. the relationship).
- On a third level, entity and relationship attributes might contain fuzzy values. This allows the imprecision to be integrated in the attributes values.

In order to explicit the potential of fuzzy database systems, the Fuzzy Enhanced Entity-Relationship model (FuzzyEER) proposed by Elmasri and Navathe [34] is shortly summarized. FuzzyEER is an extension of the Enhanced Entity-Relationship model (EER) with fuzzy semantics and notations [41]. This model firstly allows the representation of fuzzy values by means of fuzzy attributes and fuzzy degrees. Fuzzy attributes, besides the support of unknown and undefined values, enable the representation of uncertain values by the use of possibility distributions and/or similarity relations over the domain of the attribute, e.g. the unknown age of a person can be characterized by 'about 35 years old'. Fuzzy degrees, in contrast, specify an additional grade for one or several attributes (fuzzy or not) whose meaning can be a fulfillment degree, an uncertainty degree, a possibility degree, an importance degree or a membership degree. Extreme cases of fuzzy degrees occur when a grade isn't assigned to any attribute and when a grade is assigned to the whole tuple, i.e. defining a fuzzy entity, just like in the basic fuzzy relational database model. In a similar manner, an entity, a relationship or an attribute can be associated with a fuzzy degree in respect to the model. The implementation of an importance degree in a model allows, for instance, a user to view the most important part of the model according to a given importance threshold. Other constructs of the FuzzyEER model are the fuzzy weak entities, the fuzzy aggregations, the fuzzy specializations and the fuzzy constraints.

Despite the flexibility offered by fuzzy databases systems, the fuzzy classification approach proposed in this thesis does not rely on such systems. As Chap. 3 relates, information systems are mostly stored in relational databases [7] and the use of fuzzy database systems would imply the migration of the existing data. Furthermore, as the management perspectives are changing over time, the fuzzy information contained in the fuzzy databases would have to be regularly adapted. For these reasons, the chosen fuzzy classification approach works with relational databases and implements a fuzzy classification database schema (see Chap. 3). The extension of the relational database schema, which consists of meta-tables added to the system catalog, enables the definition of fuzzy classifications (see Chap. 7). Since the meta-tables are independent from the business data, changes in the fuzzy classifications' definition do not impact the original data collections.

Chapter 3
Relational Databases & Fuzzy Classification

This chapter goes into the main subject of this thesis, namely the fuzzy classification approach achieved on top of relational databases with its associated query language. Section 3.1 first motivates the strengths of combining the fuzzy classification approach with relational databases. Section 3.2 presents the different extensions of the relational model which enable a fuzzy classification, i.e. the calculation of the membership vector of the classified elements in the different classes (see Definition 2.23). Then, Sect. 3.3 introduces the fuzzy classification query language fCQL which allows the formulation of intuitive fuzzy queries with the help of linguistic variables and terms. Finally, Sect. 3.4 discusses other fuzzy approaches which combine fuzziness and relational databases.

3.1 Databases and Fuzziness

In practice, information systems are often based on very large data collections, mostly stored in relational databases [7]. Due to an information overload, it is becoming increasingly difficult to analyze these collections and to generate business decisions [33]. To address this issue, a toolkit for classification, analysis and decision support named fuzzy Classification Query Language (fCQL) has been developed [78, 79, 123]. This toolkit is a combination of relational databases and fuzzy logic. Unlike statistical data mining techniques such as cluster or regression analysis, fuzzy logic enables the use of non-numerical values and introduces the notion of linguistic variables (see Definition 2.2). Using linguistic variables and terms hides the complexity of the domain and enables a more intuitive and human-oriented querying process.

The proposed fCQL toolkit reduces the complexity of business data and extracts valuable hidden information through a fuzzy classification. The main advantage of a fuzzy classification compared to a classical one is that an element is not limited to a single class but can be assigned to several classes. Furthermore, each element has one or more membership degrees which illustrate to what extent this element belongs to the classes it has been assigned to. The notion of membership gives a much better

© Springer International Publishing Switzerland 2015
N. Werro, *Fuzzy Classification of Online Customers*, Fuzzy Management Methods,
DOI 10.1007/978-3-319-15970-6_3

description of the classified elements and also helps to reveal their potentials as well as their possible weaknesses.

The fCQL toolkit transforms fCQL queries into Structured Query Language (SQL) statements for sharp databases, thus allowing business managers to formulate and analyze unsharp queries at a linguistic level. Being an additional layer above relational database systems the proposed fuzzy classification approach guaranties a full compatibility with legacy applications. The fCQL toolkit also provides a graphical user interface to define the fuzzy classifications, meaning that the fuzzy classes, the linguistic variables and terms as well as the membership functions can be defined using a user friendly wizard (see Chap. 7).

Another important issue, considering the size and the security concern of the data collections, is that neither modification of the underlying databases nor migration of the existing data have to be undertaken. The fuzzy classification is achieved by an extension of the relational database schema in such a way that it directly operates on the underlying databases and requires no migration of the raw data. Furthermore the SQL commands as well as the transaction and recovery mechanisms offered by the Relational Database Management System (RDBMS) are still available.

In everyday business life, many examples can be found where the fuzzy classification approach would be useful. In the customer relationship management for instance, a standard classification would sharply classify customers of a company into a certain segment depending on their buying power, age and other attributes. If the client's potential of development is taken into account, the clients often cannot be classified into only one segment anymore, i.e. customer equity (see Chap. 5). Other application domains discussed in the outlook of this thesis are the portfolio analysis, the credit worthiness, the marketing mix theory and some personalization issues.

3.2 Extension of the Relational Model

The relational model has been developed by Codd in 1970 [22]. It organizes the data with the help of two-dimensional tables which are easy to display and to interpret. These tables, called relations, have a number of columns which determine the attributes of the objects to be stored. The rows of the relations, called tuples, contain the objects by assigning each attribute a determined value of the object under consideration.

The attributes assign to each entry of the relation a determined value, called tuple component, which belongs to a predefined domain [77]. Considering a set $A = A_1, \ldots, A_n$ of attributes, it concretely means that each attribute $A_j \in A$ is defined on a predefined finite domain $D(A_j)$ containing atomic values. Table 3.1 shows the tabular presentation of a relation R.

In order to avoid data redundancies and consequent data anomalies, normal forms have been developed [72, 95, 120, 121]. Five normal forms can be distinguished theoretically. In practice, most database designers concentrate on the first three normal forms which deal with the atomicity of the tuple components, the total functional

Table 3.1 Tabular representation of a relation

R						Tuple
A_1	A_2	...	A_j	...	A_n	
t_{11}	t_{12}	...	t_{1j}	...	t_{1n}	t_1
t_{21}	t_{22}	...	t_{2j}	...	t_{2n}	t_2
...	...	...	...	...	...	...
t_{m1}	t_{m2}	...	t_{mj}	...	t_{mn}	t_m

dependency of the primary keys and the transitive dependencies. Due to the respect of the normal forms, relational databases suffer from the following restrictions [139]:

- Queries to a relational database system are sharp queries. They have a dichotomic character since the criteria value and the tuple component are either identical or not. A retrieval of tuples which are more or less identical is not possible.
- The data in a relational database is precise, i.e. it is not ambiguous or unclear (with exception of null values [77]). This comes from the requirement of the classical relational data model which only accepts atomic tuple components.
- The data in a relation database is certain, i.e. the structures and relationships in a database are definitely known so that no doubt about the stored values can arise.

These restrictions clearly show that ambiguity, fuzziness, fuzzy queries as well as fuzzy classification cannot be achieved on relational databases without specific extensions of the relational model. Sections 3.2.1 and 3.2.2 therefore present the extension of the relational model by a context model and the associated context based relational algebra. The introduction of the linguistic variable concept and the calculation of the membership vector with the help of an aggregation operator are explained in the Sects. 3.2.3 and 3.2.4 respectively.

3.2.1 Database Schema with Contexts

The proposed classification approach extends the relational database schema by a context model proposed by Chen [20]. It is important to note that this extension is implemented at the database schema level while leaving unchanged the underlying databases. No migration of the raw data is therefore needed.

In the context model, each attribute is assigned a context which partitions the domain of the attribute into equivalence classes. Concretely, each attribute A_j defined on a domain $D(A_j)$ is assigned a context $C(A_j)$ [112]. Consequently, a relational database schema with contexts $R(A, C)$ is the set $A = (A_1, \ldots, A_n)$ of attributes with associated contexts $C = (C_1(A_1), \ldots, C_n(A_n))$ [47].

Throughout this chapter, a straightforward example from customer relationship management is illustrated. Furthermore, Chap. 6 provides a comprehensive classification example. For simplicity, customers are evaluated based on two attributes, turnover and behavior. The corresponding database schema with contexts *Customer ProductTable*(A, C) consists of two sets: the set A = (*Customer, Product, Turnover, Behavior*) of attributes and the set C = (C(*Customer*), C(*Product*), C(*Turnover*), C(*Behavior*)) of contexts. The turnover is a quantitative attribute measured in Euro, i.e. D(*Turnover*) = [0, 1000]. In contrast, the behavior is a qualitative attribute with the domain D(*Behavior*) = {*bad, sufficient, good, excellent*}. For the contexts of the attributes of A, the following partitions are considered:

C(*Customer*) = {{*Name of Customer*}}

C(*Product*) = {{*Product*}}

C(*Turnover*) = {[0, 499], [500, 1000]}

C(*Behavior*) = {{*bad, sufficient*}, {*good, excellent*}}

Customers and products, as they are not qualifying attributes for the classification, are both defined in one equivalence class. In contrast, the domain of the attribute turnover has been divided into two equivalence classes, the first containing the turnover from 0 to 499 €, the second with turnover from 500 to 1000 €. The attribute behavior has also two equivalence classes, i.e. 'bad' and 'sufficient' are considered to be equivalent and the same applies to 'good' and 'excellent'.

The selection of the two attributes turnover and behavior and the definition of the corresponding equivalence classes determine a two-dimensional classification space (see Fig. 3.1). The four resulting classes C_1 to C_4 can be characterized by marketing strategies such as 'Commit Customer' (C_1), 'Improve Behavior' (C_2), 'Augment Turnover' (C_3), and 'Don't Invest' (C_4).

In the proposed context model, a new form of redundancy, namely the context based redundancy, can be found.

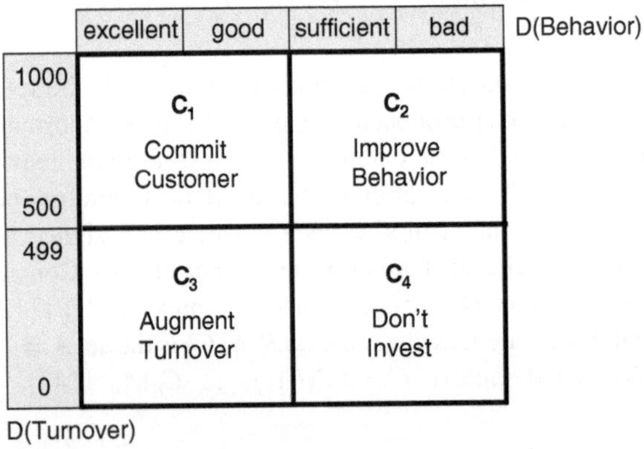

Fig. 3.1 Classification space determined by turnover and behavior

	excellent	good	sufficient	bad	D(Behavior)
1000		C_1		C_2	
		Commit Customer		Improve Behavior	
500					
499		C_3		C_4	
		Augment Turnover		Don't Invest	
0					

D(Turnover)

Definition 3.1 Two tuples t and t' of a relation r with associated schema $R(A, C)$ are context redundant regarding the corresponding set of contexts $C = (C_1(A_1), \ldots, C_n(A_n))$ if all tuple components of t_j and t'_j belong to the same equivalence class [111].

In contrast to the relational model, context redundant tuples are not identical but equivalent. Therefore, context redundant tuples do not have to be eliminated as it is the case in the relational model but they have to be combined using a merge operator.

Definition 3.2 The *merge operation* of two context redundant tuples t and t' leads to a new tuple $u = (u_1, \ldots, u_n)$ and is defined as the set theoretic union $u_j := t_j \cup t'_j$ for all tuple components u_j [111].

In the relation *CustomerProductTable-X* of Table 3.2 regarding the discussed relational database schema *CustomerProductTable(A, C)* with contexts, the tuples t_1 and t_4 as well as the tuples t_2 and t_3 are context redundant. Considering tuples t_1 and t_4, the turnover values 1000 and 600 belong to the same equivalence class; in a similar manner, so do the behavior attribute values *excellent* and *good*.

By applying the merge operator to the context redundant tuples, the relation *CustomerProductTable-X* is transformed into the context based relation *CustomerProductTable-Y* (see Table 3.3). Note that the tuples t_1 and t_4 now belong to the class C_1 (Commit Customer) whereas tuples t_2 and t_3 are classified in C_4 (Don't Invest).

Table 3.2 Database relation with context redundant tuples

CustomerProductTable-X				
Customer	Product	Turnover	Behavior	Tuple
{Smith}	{ProductC}	{1000}	{excellent}	t_1
{Ford}	{ProductA}	{400}	{sufficient}	t_2
{Miller}	{ProductB}	{50}	{bad}	t_3
{Brown}	{ProductA}	{600}	{good}	t_4

Table 3.3 Imprecise relation resulting from the merge operation

CustomerProductTable-Y				
Customer	Product	Turnover	Behavior	Class
{Smith, Brown}	{ProductC, ProductA}	{1000, 600}	{excellent, good}	C_1
{Ford, Miller}	{ProductA, ProductB}	{400, 50}	{sufficient, bad}	C_4

3.2.2 *Context Based Relational Algebra*

The classical relational algebra consists of three set operators (the set theoretic union, the set theoretic difference and the cartesian product) and two relational operators (the projection π and the selection σ) [77]. These five operators form the minimal set of operators since all other set and relational operators can be expressed by a combination of them. A relational query language is relationally complete if it has, at least, the power of the five basic algebraic operators [22]. Well-known relational query languages such as SQL are relational complete [19].

The context based relational algebra is an extension of the classical relational algebra which uses relational database schemas $R(A, C)$ with contexts. For a given set of querying attributes QA and a set of corresponding querying contexts QC there exists a context based union, difference and cartesian product as well as a context based projection Π and selection Σ [35, 111]. To illustrate the context based relational algebra, the projection and selection operators are discussed.

The context based projection, just as the classical one, reduces the columns of a relation to the specified attributes. However, the treatment of the redundant tuples is different as context redundant tuples are merged instead of being eliminated (see Definitions 3.1 and 3.2).

Definition 3.3 For a relation r of a database schema $R(A, C)$ with contexts and a schema $R(QA, QC)$ with a query set of attributes $QA \subset A$ and a set of associated query contexts QC there exists a *context based projection* $\Pi[QA, QC](r)$. The projection of r regarding $R(QA, QC)$ is a relation qr of $R(QA, QC)$. The tuples in qr are calculated by reducing the tuple components of r to the set of querying attributes QA. Context redundant tuples regarding QC are merged by the merge operator [35, 111].

The selection operator restricts a relation to a subset of tuples based on a selection condition. The context based selection extends the classical one by replacing the notion of equality by the notion of equivalence for the assessment of the selection criteria. The context based projection also handles the context redundant tuples accordingly.

Definition 3.4 For a relation r of a database schema $R(A, C)$ with contexts and a schema $R(QA, QC)$ with a query set of attributes $QA \subset A$ and a set of associated query contexts QC there exists a *context based selection* $\Sigma[\beta(QA, QC)](r)$ where $\beta(QA, QC)$ is a boolean condition for values of the attribute set QA and corresponding query contexts QC [35].

In order to explicit the definition of the context based projection and selection, consider the *CustomerProductTable* in Table 3.4 which contains an extended sample of customers and products as well as the following querying attributes and contexts:

Table 3.4 Customer and product relation with contexts

CustomerProductTable			
Customer	Product	Turnover	Behavior
{Smith}	{ProductC}	{1000}	{excellent}
{Smith}	{ProductB}	{700}	{good}
{Ford}	{ProductA}	{400}	{sufficient}
{Ford}	{ProductC}	{600}	{excellent}
{Miller}	{ProductB}	{50}	{bad}
{Miller}	{ProductC}	{300}	{sufficient}
{Brown}	{ProductA}	{600}	{good}
{Brown}	{ProductB}	{400}	{excellent}

$QC(Customer) = \{\{Name\ of\ Customer\}\}$

$QC(Product) = PC(Product)$

$QC(Turnover) = \{[0, 499], [500, 1000]\}$

$QC(Behavior) = \{\{bad,\ sufficient\}, \{good,\ excellent\}\}$

The querying contexts QC slightly differ from the contexts defined in Sect. 3.2.1 in the sense that the precise contexts (PC) of the attribute *Product* are now considered, i.e. each product is in an own equivalence class. This modification enables the selection of customers based on a specific product. For instance, if customers who bought product C have to be assessed, a context based projection and selection can be expressed as follows:

CustomerRatingTable $- X :=$

$\Pi[Customer, Turnover, Behavior, QC()]$

$(\Sigma[Product_{\sim QC(Product)}ProductC](CustomerProductTable))$

The execution of this query is performed in two steps: first the context based selection selects all customers having bought product C, then the context based projection reduces the temporary relation to the specified attributes. Note that both operators eliminate the context redundant tuples. The result of the query, which is the imprecise relation *CustomerRatingTable* $- X$ shown in Table 3.5, leads to a sharp classification as every customer belongs to exactly one class.

Querying the context based relation *CustomerProductTable* not necessarily results in a sharp classification of customers. If, for instance, customers are to be rated independently of the purchased products, the following context based projection can be formulated:

CustomerRatingTable $- Y :=$

$\Pi[Customer, Turnover, Behavior, QC()](CustomerProductTable)$

Table 3.5 Sharp classification of customers regarding product C

CustomerRatingTable-X			
Customer	Turnover	Behavior	Class
{Smith, Ford}	{1000, 600}	{excellent}	C_1
{Miller}	{300}	{sufficient}	C_4

Table 3.6 Unsharp classification of customers

CustomerRatingTable-Y			
Customer	Turnover	Behavior	Class
{Smith, Ford, Brown}	{1000, 700, 600}	{excellent, good}	C_1
{Brown}	{400}	{excellent}	C_3
{Ford, Miller}	{400, 300, 50}	{sufficient, bad}	C_4

The result of this query, the context based relation *CustomerRatingTable − Y* illustrated in Table 3.6, produces an unsharp classification since some customers belong to several classes at the same time. Although customers Smith and Miller are still classified sharply, customers Brown and Ford belong to two different equivalence classes at the same time.

This simple example demonstrates that querying a relation with contexts may result in a non sharp classification where the classified objects can be assigned to more than one equivalence class. This leads to an uncertainty regarding the interpretation of the results. In fact, it is necessary to know to what extent an element belongs to a certain class, i.e. to perform a fuzzy classification where membership degrees indicate the belonging of the elements to the different classes.

3.2.3 Fuzzy Classification Database Schema

In this subsection, the context model is extended by defining a fuzzy classification database schema. This extension enables the determination of the elements' membership degrees in the different classes, i.e. the calculation of the membership vector, and also permits a querying process on a linguistic level, i.e. without numerical values (see Sect. 3.3).

To derive fuzzy classes from sharp contexts, each qualifying attribute is considered as a linguistic variable (see Definition 2.2) and verbal terms are assigned to each equivalence class. The verbal terms of the linguistic variables are words or word combinations which describe the equivalence classes. This leads to a fuzzy classification database schema $R(A, C, X, T)$ which is a database schema with a set

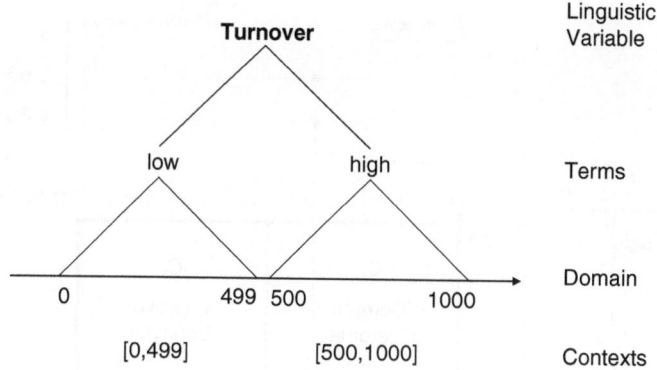

Fig. 3.2 Concept of linguistic variable applied to the context based model

of attributes A, a set of associated contexts C, a set of linguistic variables X, and a set of corresponding terms T [102]. Each linguistic variable X_i has an associated set of terms $T(X_i) := T_1, \ldots, T_k$. In other words, a fuzzy classification database schema is a database schema with contexts, extended by linguistic variables and terms.

As an example, consider the attribute turnover as a linguistic variable (see Fig. 3.2). The definition range of the linguistic variable is the domain of the attribute turnover. The linguistic variable holds the set of terms $T(Turnover) = \{low, high\}$ describing the two equivalence classes [0, 499] and [500, 1000].

With the concept of linguistic variables, the equivalence classes are described more intuitively. In addition, every term of a linguistic variable represents a fuzzy set that is determined by a membership function μ over the domain of the corresponding attribute [134, 135, 136].

In Fig. 3.3, the terms *low* and *high* of the linguistic variable turnover are associated with the membership functions μ_{high} and μ_{low} which are defined on the entire domain of the attribute turnover. It is therefore possible for a specific turnover value, for instance the 510 € turnover of customer Brown, to be at the same time judged as 'high' and 'low'. This means that the membership degrees of Brown in the equivalence classes 'high' and 'low' are respectively 0.52 and 0.48 (compare with Fig. 3.3). The same is true for the linguistic variable behavior with its terms *attractive* and *non attractive*.

The definition of the vague terms 'high' and 'low' with corresponding membership functions leads to a fuzzy partition of the domain of the attribute turnover. This fuzzy partition has an important outcome, it implies the disappearance of the classes' sharp borders replaced by continuous transitions between the different classes. A customer can consequently belong to more than one class at the same time and his membership degrees in different classes can be calculated.

As turnover is a numerical attribute in the interval [0, 1000], the membership functions of terms belonging to the linguistic variable turnover are continuous. In contrast, behavior being a qualitative attribute, the terms of the linguistic variable behavior

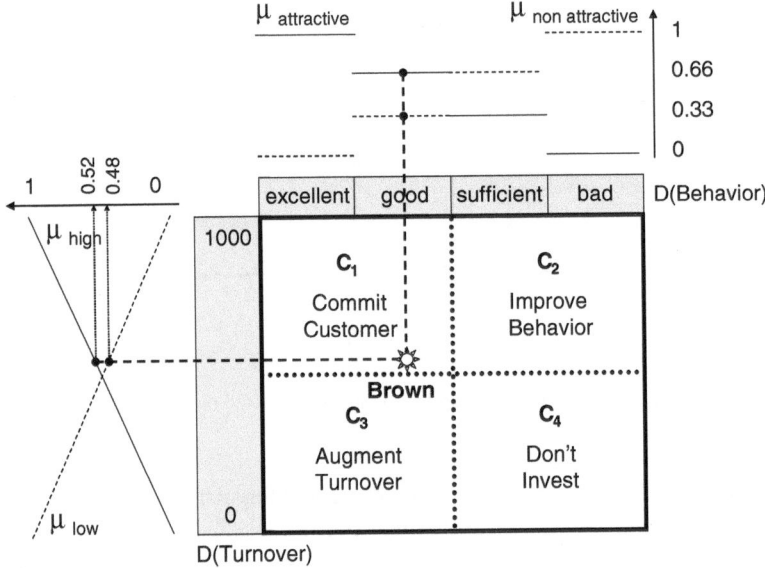

Fig. 3.3   Fuzzy classification with membership functions

are associated with discrete membership functions, i.e. each term corresponds to a discrete value (see Fig. 3.3).

3.2.4 Aggregation Operator

The membership degree $M(O_i|C_k)$ of an object O_i in a class C_k can be calculated by an aggregation over all terms of the linguistic variables which define the class. The class C_1 for instance is described by the terms 'high' and 'attractive' of the linguistic variables turnover and behavior respectively. The membership degree of an object in class C_1 is then the aggregation of the corresponding values of the membership functions μ_{high} and $\mu_{attractive}$.

There exists quite a number of operators which calculate the aggregation of membership values (see Appendix A) [139]. Well known aggregation operators, originally proposed by Zadeh [131], are the minimum and maximum operators (see Definitions 2.15 and 2.16). On the one hand, using the minimum operator for the aggregation distinguishes the smallest membership function values from all others meaning that no compensation effect is performed. On the other hand, the use of the maximum operator, which considers the largest membership degrees, implies a full compensation between the membership functions. When classifying a customer with a high turnover and a non attractive behavior, the minimum (resp. maximum) operator would judge this customer only by his behavior (resp. turnover). Following

Table 3.7 Customer relation with turnover and behavior information

CustomerTable		
Customer	Turnover	Behavior
Brown	{510}	{good}
Smith	{1000}	{excellent}
Miller	{50}	{bad}
Ford	{490}	{sufficient}

Zimmermann and Zysno [140], managerial decisions rarely have any compensation nor imply a complete compensation, but almost always show some degree of compensation. Consequently, a manager would normally weigh up both attributes turnover and behavior. This human judgment, i.e. the compensation of non attractive behavior by high turnover, can be achieved by using the γ-operator (see Definition 2.21). The γ-operator was empirically tested and proven to be more appropriate in human decision making than other averaging or compensatory operators [140].

In order to illustrate the classes belonging calculation, consider the simplified customer relation *CustomerTable* which does not contain references to products (see Table 3.7) and customer Brown who will be analyzed in details.

In a first step, the raw membership degrees of Brown in the classes C_1 to C_4 are calculated using the γ-operator and a γ-value of 0.5:

$$M_{raw}(Brown|C_1) = (\mu_{high}(510) \times \mu_{attractive}(good))^{0.5} \times$$
$$(1 - (1 - \mu_{high}(510)) \times (1 - \mu_{attractive}(good)))^{0.5} =$$
$$(0.52 \times 0.66)^{0.5} \times (1 - 0.48 \times 0.34)^{0.5} =$$
$$(0.3432)^{0.5} \times (0.8368)^{0.5} \approx 0.5858 \times 0.9148 \approx 0.5359$$

$$M_{raw}(Brown|C_2) = (\mu_{high}(510) \times \mu_{non\,attractive}(good))^{0.5} \times$$
$$(1 - (1 - \mu_{high}(510)) \times (1 - \mu_{non\,attractive}(good)))^{0.5} =$$
$$(0.52 \times 0.33)^{0.5} \times (1 - 0.48 \times 0.67)^{0.5} =$$
$$(0.1716)^{0.5} \times (0.6784)^{0.5} \approx 0.4142 \times 0.8237 \approx 0.3412$$

$$M_{raw}(Brown|C_3) = (\mu_{low}(510) \times \mu_{attractive}(good))^{0.5} \times$$
$$(1 - (1 - \mu_{low}(510)) \times (1 - \mu_{attractive}(good)))^{0.5} =$$
$$(0.48 \times 0.66)^{0.5} \times (1 - 0.52 \times 0.34)^{0.5} =$$
$$(0.3168)^{0.5} \times (0.8232)^{0.5} \approx 0.5628 \times 0.9073 \approx 0.5107$$

$$M_{raw}(Brown|C_4) = (\mu_{low}(510) \times \mu_{non\,attractive}(good))^{0.5} \times$$
$$(1 - (1 - \mu_{low}(510)) \times (1 - \mu_{non\,attractive}(good)))^{0.5} =$$

$$(0.48 \times 0.33)^{0.5} \times (1 - 0.52 \times 0.67)^{0.5} =$$
$$(0.1584)^{0.5} \times (0.6516)^{0.5} \approx 0.3980 \times 0.8072 \approx 0.3213$$

The raw membership degrees are only meaningful when elements are evaluated within a single class. If two or more classes are considered, the raw membership degrees have to be normalized, meaning that the total belonging to all classes sums up to 1.00. In the case of customer Brown, the total belonging which is the cardinality of Brown in classes C_1 to C_4 (see Definition 2.6) is:

$$Card(Brown|C_i) = \sum_{i=1}^{4} M_{raw}(Brown|C_i) \approx 1.7090$$

Then, the normalization takes place in order to derive the final membership degrees in the four classes:

$$M_{final}(Brown|C_1) = M_{raw}(Brown|C_1) \div Card(Brown|C_i) =$$
$$0.5359 \div 1.7090 \approx 0.3136 \approx 0.31$$
$$M_{final}(Brown|C_2) = M_{raw}(Brown|C_2) \div Card(Brown|C_i) =$$
$$0.3412 \div 1.7090 \approx 0.1996 \approx 0.20$$
$$M_{final}(Brown|C_3) = M_{raw}(Brown|C_3) \div Card(Brown|C_i) =$$
$$0.5107 \div 1.7090 \approx 0.2988 \approx 0.30$$
$$M_{final}(Brown|C_4) = M_{raw}(Brown|C_4) \div Card(Brown|C_i) =$$
$$0.3213 \div 1.7090 \approx 0.1880 \approx 0.19$$

In a similar manner, the class membership degrees of customers Smith, Miller and Ford can be calculated and the result is a context redundant-free relation *CustomerRatingTable* − *Z* shown in Table 3.8. Applying the γ-operator has produced a fuzzy relation, i.e. the tuple components are now fuzzy sets. The customers can belong to different classes and these belongings are weighted by membership degrees. As shown in the calculation details, customer Brown belongs to all four classes at the same time with different membership degrees; the same applies to customer Ford. It is important to note that customers can still be classified sharply, for instance customers Smith and Miller fully belong to C_1 and C_4 respectively. The number of sharply classified customers strongly depends on the definition of the terms' membership functions, however customers who are located in a corner of the classification space are always classified sharply.

In order to improve the readability of the results, the fCQL toolkit presented in Chap. 7 always returns a customer-centered results table as shown in Table 3.9. It contains columns for the attributes specified in the projection and one column for each resulting class. Each customer is then represented by a tuple and the belonging to the different classes appears in their corresponding columns.

Table 3.8 Fuzzy classification with membership degrees

CustomerRatingTable-Z			
Customer	Turnover	Behavior	Class
{(Smith, 1.00), (Brown, 0.31), (Ford, 0.19)}	high	attractive	C_1
{(Brown, 0.20), (Ford, 0.30)}	high	non attractive	C_2
{(Brown, 0.30), (Ford, 0.20)}	low	attractive	C_3
{(Brown, 0.19), (Ford, 0.31), (Miller, 1.00)}	low	non attractive	C_4

Table 3.9 fCQL generated results table

ResultsTable						
Customer	Turnover	Behavior	C_1	C_2	C_3	C_4
Brown	510	good	0.31	0.20	0.30	0.19
Smith	1000	excellent	1	0	0	0
Miller	50	bad	0	0	0	1
Ford	490	sufficient	0.19	0.30	0.20	0.31

Also, for discussion purpose, the following textual notation is used in the remainder of this thesis to depict the customers' belonging to the classes:

Brown $(C_1:0.31, C_2:0.20, C_3:0.30, C_4:0.19)$
Smith $(C_1:1, C_2:0, C_3:0, C_4:0)$
Miller $(C_1:0, C_2:0, C_3:0, C_4:1)$
Ford $(C_1:0.19, C_2:0.30, C_3:0.20, C_4:0.31)$

To conclude this section, note that the selection of qualifying attributes, the introduction of equivalence classes, the definition of linguistic variables and terms and the choice of membership functions are important design issues [79]. Database architects and specialists from different departments have to work together in order to devise an adequate fuzzy classification.

3.3 Fuzzy Classification Query Language fCQL

Having defined a fuzzy classification database schema with linguistic variables and terms, the user needs a way to query it. The fuzzy Classification Query Language fCQL, implemented in the fCQL toolkit (see Chap. 7), allows the user to formulate unsharp classification queries by taking advantages of the predefined linguistic variables and terms as well as the classes of the classification space.

This section first discusses the principles of the fuzzy classification query language in Sect. 3.3.1, then concretely introduces the syntax of fCQL in Sect. 3.3.2 and finally depicts some fCQL query examples with their results in Sect. 3.3.3.

3.3.1 fCQL Principles

The fCQL toolkit, shortly mentioned in Sect. 3.1, is a classification, analysis and decision support tool. It allows business managers to query predefined fuzzy classifications in relational databases in order to improve the quality of their decisions [78]. In contrast to other fuzzy query languages (see Sect. 3.4), the user does not need to deal with a fuzzy SQL or with fuzzy predicates which could lead to varying semantics and different interpretations of the original data collection [35].

The use of a predefined fuzzy classification database schema can also be justified by the fact that only experts of the domain can wisely determine the pertinent attributes for the classification, decide the appropriate equivalence classes and finally design the membership functions. Consequently, the proposed approach hides the complexity of the domain from the end users and allows them to focus on the querying process [123].

From the user's point of view, fCQL can be seen as a human-oriented query language as it works on a linguistic level, i.e. without numerical values, through the use of predefined linguistic variables and their associated verbal terms. The user can therefore easily formulate classification queries as they are intuitive, i.e. the meaning of the queries is linguistically expressed.

Having a predefined set of linguistic variables and terms is also a way of establishing a common terminology across and within the departments of a company [84]. This is particularly useful in large enterprises where the technical department, comprising for instance a data mining team, and the marketing department are often siloed and usually have a totally different terminology.

3.3.2 fCQL Syntax

The Structured Query Language SQL is the standard for defining and querying relational databases [19]. It would have been possible to use the view concept of SQL in order to define classes. Also, access to sets could have been formulated with the inclusion predicate of SQL. In this case, however, the user of a relational database with fuzzy classification would not be able to retrieve membership values and to evaluate the query results thoroughly. An extension of the SQL language is therefore necessary. The proposed extension is the fuzzy Classification Query Language fCQL, originally described by Schindler [102] (see Appendix B) and slightly improved with the addition of the SQL *where* clause as well as a new *alpha* clause:

classify   AttributeList
from   RelationName
[where   SelectionCondition]
[with   ClassificationCondition]
[alpha   AlphaCondition]

The classification language fCQL has been designed in the spirit of SQL. Instead of specifying the attribute list in the select clause, the attributes to be classified are given in the *classify* clause. The *from* clause specifies the considered relation, just as in SQL. It is also possible to classify a subset of objects by specifying a selection condition in the optional *where* clause. Most important, a *with* and an *alpha* clauses are introduced. The *with* clause lets the user specify a classification predicate whereas the *alpha* clause allows him to enter selection conditions on the fuzzy classification results based on α-cuts (see Definition 2.9).

It is in the *with* (resp. *alpha*) clause that the user can take advantage of the predefined linguistic variables and their associated verbal terms (resp. of the predefined classes and concepts). By combining linguistic variables and terms with keywords, the user can formulate classification conditions. Similarly, the user can generate alpha conditions by specifying restrictions on the final classes and/or on the predefined concepts (the notion of concept is introduced in Chap. 5). Note that the *where*, *with* and *alpha* conditions are not compulsory. A syntax diagram showing the creation of fCQL queries is given in Fig. 3.4. Boxes with round corners contain fCQL keywords and rectangles are elements of the relation or fuzzy classification definition. The original grammar of fCQL can be found in Appendix B.

3.3.3 fCQL Query Examples

As a first example, consider the fCQL query that performs a fuzzy classification of all customers in the four classes C_1 to C_4:

classify   Customer, Turnover, Behavior
from   CustomerTable

This query classifies all customers of the relation *CustomerTable* (see Table 3.7) and produces the already discussed results table of Table 3.9. It is also possible to dynamically ask for new classes. This is done by querying specific terms of the linguistic variables. For instance, customers with behavior problems can be rated by the following query:

classify   Customer, Behavior
from   CustomerTable
with   Behavior **is** non attractive

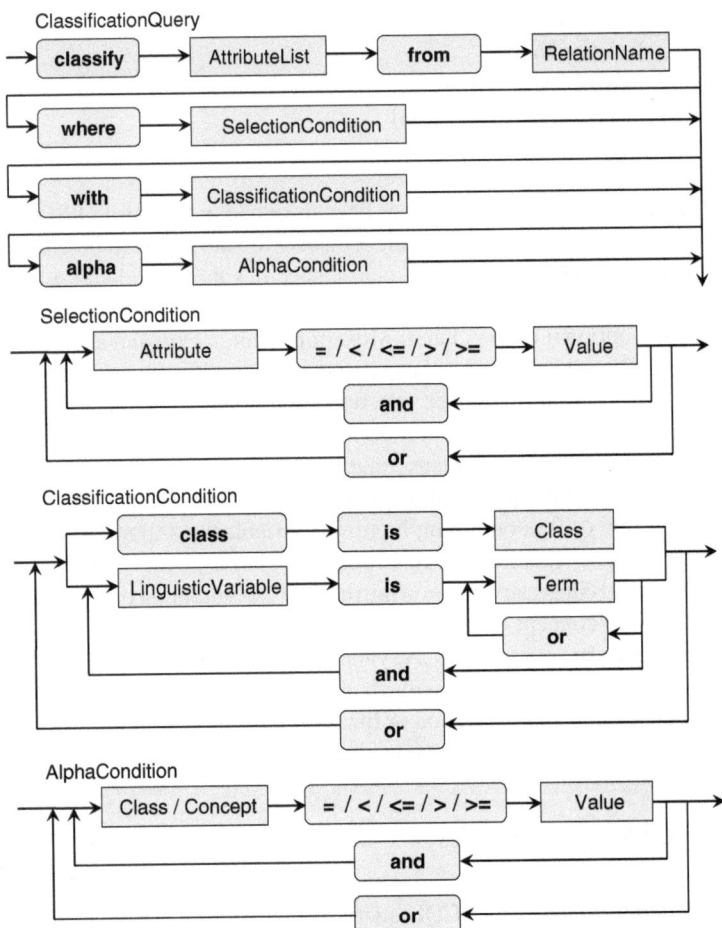

Fig. 3.4 Syntax diagram of fCQL

   This classification query evaluates customers based only on their behavior in the
equivalence class 'non attractive'. As only one linguistic variable and term are considered, no aggregation takes place and the membership values are directly derived from
the membership function $\mu_{non\,attractive}$. The customers are consequently classified in
a new class (see Table 3.10).

   In contrast to the predefined classes C_1 to C_4 which were given a name (resp.
a semantics), the dynamically created classes are not labelled, therefore their name
is the concatenation of the names of the linguistic variable and term as well as the
chosen connectors, if any. Querying new classes requires a good knowledge of the
fuzzy classification definition as the semantics of these classes has to be carefully
interpreted to avoid misleading conclusions.

Table 3.10 Classification results of customers in a new class

ResultsTable		
Customer	Behavior	Behavior/non attractive
Brown	good	0.33
Smith	excellent	0
Miller	bad	1
Ford	sufficient	0.66

Table 3.11 Classification results of customers in a single predefined class

CustomerRatingTable-Z			
Customer	Turnover	Behavior	Class
{(Smith, 1.00), (Brown, 0.31), (Ford, 0.19)}	high	attractive	C_1
{(Brown, 0.20), (Ford, 0.30)}	high	non attractive	C_2
{(Brown, 0.30), (Ford, 0.20)}	low	attractive	C_3
{(Brown, 0.19), (Ford, 0.31), (Miller, 1.00)}	low	non attractive	C_4

Going one step further, a classification query can be formulated using two linguistic variables and terms:

classify Customer, Turnover, Behavior
from CustomerTable
with Turnover **is** high **and** Behavior **is** attractive

In fact this query returns the predefined class C_1 with semantics 'Commit Customer'. Remember that the class C_1 has been defined as the aggregation of the terms 'high' and 'attractive'. A straightforward alternative to retrieve the same results is to specify the class in the classification predicate:

classify Customer, Turnover, Behavior
from CustomerTable
with **class is** Commit Customer

The classification takes place again in a single class but due to the selection of two linguistic variables and terms, an aggregation of the corresponding terms is required. As discussed in Sect. 3.2.4, the raw and final membership degrees can be distinguished. In this case, no normalization is required as only one class is considered. The figures of the results table shown in Table 3.11 are therefore different from those of the class C_1 in Table 3.9.

Table 3.12 Classification results of customers with a selection condition

ResultsTable			
Customer	Turnover	Behavior	Commit Customer
Smith	1000	excellent	1

Now, the user might want to analyze only the best customers in class C_1. Such a selection can be achieved by formulating a restriction on the attributes values in the *where* clause:

classify   Customer, Turnover, Behavior
from   CustomerTable
where   Turnover $\geq$ 750 **and** Behavior $=$ excellent
with   **class is** Commit Customer

A more pertinent way of selecting the best customers is to use the *alpha* clause which allows the user to apply restrictions on the classification results. This way, the user can benefit from the class definitions and their associated semantics:

classify   Customer, Turnover, Behavior
from   CustomerTable
with   **class is** Commit Customer
alpha   Commit Customer $\geq$ 0.8

The last two fCQL queries produce identical results as shown in Table 3.12. For both queries, only customer Smith is selected since he is the only one to comply with the definition of 'best customers' expressed in the *where* and *alpha* clauses.

In this simple example, specifying linguistic variables and terms in the *with* clause is straightforward. However, if customers were to be classified on three or more attributes, the capability of fCQL for a multi-dimensional classification space is increased. Therefore, fuzzy queries can also be seen as an extension of the classical slicing and dicing operators on a data cube [123].

3.4 Other Fuzzy Query Languages

The fuzzy classification query language fCQL presented in Sect. 3.3 uses fuzzy logic to achieve a fuzzy classification but there exist a number of fuzzy query languages which aim to formulate general fuzzy queries against classical database systems. Unlike fuzzy database systems (see Sect. 2.5.5), relational databases only contain precise and certain data. The aim of the fuzzy query languages is therefore to make the querying of regular databases more powerful by allowing imprecise queries to be formulated. The results of such queries are no more crisp, i.e. an element satisfies or not a condition, but fuzzy, i.e. to what extent this element satisfies the condition. It therefore allows a qualitative distinction between the selected elements. Furthermore, the introduction of imprecise conditions improves the querying process when the user is not able to sharply define his needs or if an interpretation margin in the query is needed [10].

Following Kacprzyk [60], the first approach to fuzzy querying a DBMS has been proposed in 1977 by Tahani [115]. By allowing both boolean and fuzzy predicates in the queries, matching degrees of ordinary relations' tuples can be computed defining the concept of fuzzy relations. In this approach, fuzzy predicates are formed of fuzzy terms (e.g. 'young') defined as linguistic variables on the domain of the considered attribute. The predicates can then be combined using the negation and the logical connectives 'and' and 'or'.

In 1991 Takahashi developed the Fuzzy Query Language (FQL), a fuzzy query language which is an extension of the relational domain calculus [116]. FQL queries allow a natural determination method of matching degrees of domain values allowing a human-oriented interface to relational databases. This language is able to represent the five types of fuzzy propositions distinguished in the meaning representation language PRUF proposed by Zadeh [138]. Another extension of the domain calculus which allows the querying of relational as well as fuzzy relational databases has been proposed by Galindo et al. in 1999 [39].

The rest of this section presents into more details three of the major fuzzy query languages for relational databases, namely SQLf in Sect. 3.4.1, FQUERY in Sect. 3.4.2 and FSQL in Sect. 3.4.3.

3.4.1 SQLf

Since 1988, Bosc et al. have published many contributions on SQLf which is a fuzzification of SQL allowing flexible queries interpreted in the framework of the fuzzy set theory [9, 10, 11, 12, 13, 14]. "The objective of flexible queries is to support preferences and to provide users with results which are ranked according to their adequation with respect to the query" [12]. An important aspect of SQLf is its similarity to SQL on the syntax level as well as on the query equivalence level. Complex SQL queries involving nesting and set-oriented operators can have several equivalent formulations. Bosc and Pivert showed that these equivalences are still valid using an imprecise querying context [13].

In order to integrate fuzzy querying capabilities, extensions of single and nested SQL blocks can be considered. Considering the SQL base block, the fuzziness is integrated at two levels:

select   [n | t |n, t] <attributes>
from   <relations>
where   <fuzzycondition>

First, a fuzzy condition combining boolean and fuzzy predicates can be specified in the 'where' clause. Applying a fuzzy condition leads to a fuzzy relation so that, in order to retrieve a classical (non fuzzy) relation, a criterion has to be specified for selecting the most suitable elements of the fuzzy relation. Therefore, the user is given an output regulation mechanism by specifying a number of desired responses n and/or a threshold value t ∈ [0,1] in the 'select' clause.

Different fuzzy predicates can be distinguished: atomic predicates like 'young' and 'about 35' which are represented by a membership function and modified predicates which are atomic predicate combined with a unary operator, e.g. the negation or linguistic modifiers. Furthermore, compound predicates can be obtained by the use of n-ary operators, e.g. intersection, union or means.

Additional possibilities are enabled by using nested SQL blocks (subqueries) and linguistic quantifiers. Just as SQL enables the statement of conditions to sets of tuples in the 'having' clause, SQLf enables the specification of fuzzy conditions. SQLf furthermore allows fuzzy quantifiers to be included in the 'having' clause:

select	[n	t |n, t] <attributes>
from	<relations>
group by	<attributes>
having	<quantified proposition>

Considering the relations EMPLOYEE(id, name, salary, job, age, city, department) and DEPARTMENT(id, manager, budget, location) Bosc and Pivert [12] propose the following example:

select	10 department
from	EMPLOYEE
group by	department
having	most-of(age ="young") **are** "well-paid"

3.4.2 FQUERY

Another fuzzy query language whose aim is to enable a more intelligent and human consistent information retrieval is FQUERY proposed by Kacprzyk and Zadrozny [61]. It allows the user to formulate fuzzy or imprecise descriptions as well as linguistic quantifiers in the querying process. FQUERY is available for several widely used DBMSs including the Microsoft Access DBMS (as an add-on). The FQUERY language extends SQL by allowing fuzzy terms:

select <list of fields>
from <list of tables>
where {most | almost all | etc.} $cond_{11}$ **and** ... **and** $cond_{1k}$ **or**
...
$cond_{n1}$ **and** ... **and** $cond_{nm}$

Each condition is of the form 'NumericalAttribute is <fuzzy value>' or 'NumericalAttribute$_1$ <fuzzy relation> NumericalAttribute$_2$'. In order to calculate the compatibility degree of a query, the numerical fields to be queried must be declared and the lower and upper interval limits must be given. Then, fuzzy values and relations are represented with the help of a trapezoidal membership function in the interval $[-10, 10]$ on which the attributes' values are mapped. Examples of fuzzy queries are 'salary is *large*' and 'Amount in Stock *is much greater than* Amount on Orders'. A key feature of FQUERY is the use of linguistic quantifiers (most, almost all, ...) making queries more flexible. Linguistic quantifiers are defined as a fuzzy set in the interval [0, 10] and allow the user to specify how many atomic conditions and sub-conditions have to be fulfilled in a sub-condition and condition. The grammar of FQUERY can be found in Appendix B.

In 1986, Kacprzyk and Zadrozny extended the grammar of FQUERY to support Ordered Weighted Averaging (OWA) operators with linguistic quantifiers in order to have more convenient and flexible means to aggregate partial matching degrees. They also added the support of non numerical attributes with single-valued and multi-valued attributes with respect to fuzzy set constants [62]. In 2000 FQUERY has been improved to enable a fuzzy querying over the Internet via a web browser [63]. In 2001, the FQUERY grammar has been extended to integrate vertical quantified queries by adding a 'group by' and a 'having' clause [64]. The 'group by' clause allows the user to get aggregated results for the specified groups taking into account that the 'having' clause can also take advantage of linguistic quantifiers. Note that the vertical quantified queries have been suggested by Bosc et al. [10] and have been first implemented in SQLf (see Sect. 3.4.1).

3.4.3 FSQL

The Fuzzy SQL (FSQL) language proposed in 1998 by Galindo et al. [40, 41] also extends the SQL language to allow flexible queries. Despite the fact that this fuzzy query language has been designed to work with a fuzzy relational database system allowing the storage of vague information, it still relaxes the querying process of precisely stored information.

Concretely, FSQL offers the following possibilities:

- The definition of linguistic labels for attributes based on an ordered or non-ordered domain (i.e. the concept of linguistic variables). Labels for attributes having an ordered domain are defined by means of trapezoidal possibility distributions as those based on a non-ordered domain use a similarity relation.

- The use of fuzzy comparators based on possibility as well as necessity measures. These operators are Fuzzy Equal, Fuzzy Different From, Fuzzy Greater/Less Than, Fuzzy Greater/Less or Equal, Much Greater/ Less Than.
- The specification of a fulfillment threshold $\tau \in [0,1]$ which specifies the minimum degree the considered elements have to match. The threshold can be determined using the keyword THOLD or by using traditional comparators (e.g. $=$, $<$, $\leq$, $>$, $\geq$). Note that a qualifier can also be used as a threshold.
- The utilization of absolute and relative quantifiers also defined by trapezoidal membership function. Fuzzy quantifiers are permitted in the WHERE and in the HAVING clauses.
- The usage of fuzzy constants which are either user-defined constants or system constants (e.g. NULL, UNKNOWN or UNDEFINED).

The syntax of FSQL is too complex to explicit it here, for more details refer to Galindo et al. [41]. FSQL is the most complete language since it does not only refer to the SELECT statement as do the other fuzzy query languages presented in this section, but also considers the DML statements (Data Manipulation Language, i.e. INSERT, DELETE and UPDATE) and the DDL statements (Data Definition Language, i.e. CREATE, DROP AND ALTER). Also when taking into account only the SELECT statement, FSQL shows more flexibility than the other candidates, perhaps at the price of more complex queries. FSQL represents therefore the state of the art in fuzzy database modeling and querying.

Part II
Customer Perspective

Chapter 4
Customer Relationship Management

This chapter has the objective of providing the reader with a basic understanding of the customer relationship management (CRM) and its implications as the next chapters rely on it. For this sake, Sect. 4.1 first depicts the general concept of CRM, then Sect. 4.2 discusses the CRM theoretical constructs whereas Sect. 4.3 looks more closely at the architecture and the functionalities of CRM systems. Section 4.4 finally clarifies the use and the pertinence of the fuzzy classification approach in the CRM framework.

4.1 CRM Concept

In recent years, CRM has established itself in the practice [51]. CRM is a customer oriented strategy which allows enterprises to better manage their customers by developing profitable and long term customer relationships with the aim of improving their financial performance. For Diller [27], CRM is a synonym of relationship marketing, i.e. a strategic marketing concept in which the market success is achieved by a systematic analysis, planning, organization and control of the individual customer relationships. For other authors [42, 51, 53, 70], CRM is more than a new marketing strategy, it consists of strategies and technologies which enable companies to offer their customers the products and services they are looking for. More precisely, CRM is the ability of identifying, acquiring and retaining the most profitable customers in order to augment the enterprise's turnover and profits [70]. There exist many different definitions in the CRM literature. In this work the definition proposed by Hippner and Wilde [53] in Definition 4.1 is considered.

Definition 4.1 *Customer Relationship Management* is a customer oriented company's strategy which, with the help of modern information and communication technologies, aims at building and consolidating long term profitable customer relationships by means of global and individual marketing, sales and after-sales concepts.

© Springer International Publishing Switzerland 2015
N. Werro, *Fuzzy Classification of Online Customers*, Fuzzy Management Methods,
DOI 10.1007/978-3-319-15970-6_4

Based on this definition, Hippner distinguishes two central CRM components [51]:

- *Customer oriented strategy*: CRM is first of all a customer oriented strategy implying a reorganization of the business processes and of the responsibilities on the customers.
- *Integrated information systems* (*CRM systems*): In order to have a global image of the customers ('one face of customer') and a consistent communication with the customers ('one face to customer'), CRM requires information systems to centralize all customer related information and to synchronize all the communication channels.

These two aspects clearly show the main drivers of the CRM framework: on the one side, the theoretical perspective which is supported by an abundant literature and, on the other side, the technological perspective encompassing a large number of CRM systems for the storage and the analysis of the customer data as well as the automation of customer related processes. In order to be effective, CRM should rely on both aspects. In reality, however, CRM is often reduced to its technological component or, in a similar manner, many theoretical constructs do not have a concrete realization [51].

Figure 4.1 depicts a CRM success chain with four phases. The first phase aims at developing a customer oriented strategy. This strategy encompasses the firm's strategic objectives, the customer oriented management concepts and the multi channel management. The second phase is the reorganization of the company's structure and processes based on the new customer oriented strategy. This phase optimizes the customer related processes, their support by CRM systems as well as the planning and the control of the company's reorganization with the help of a change management and a CRM project management. The third phase first has an impact on the customer base since the customer orientation tries to modify the customers' opinion and behavior, namely the customer satisfaction, the customer loyalty and the customer retention. It is in the fourth phase that the financial success can be measured based on the quantity, on the quality and on the duration of the customer relationships that have been created. All this factors positively impact the customer equity (see Sect. 4.2.1).

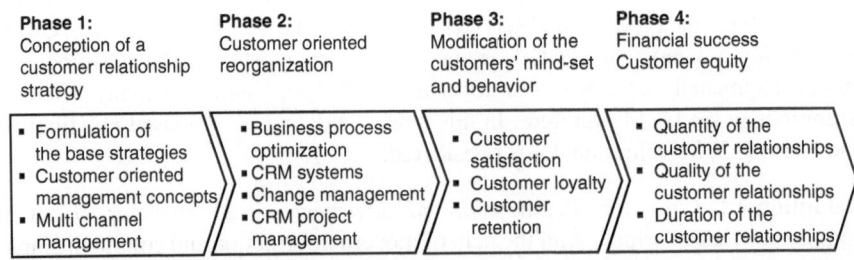

Fig. 4.1 CRM success chain (adapted from [51])

Based on the customer life cycle shown in Fig. 4.2, three strategic objectives can be distinguished in order to develop and maintain profitable customer relationships [90]:

- *Customer acquisition (or recruitment)*: The first objective is the broadening of the customer base by building new relationships. The acquisition of new customers has to be done in a differentiated way, meaning that not all customers are equally worthy to have since the firm might not be able to suit their needs or some customers are likely to be unprofitable. Company should therefore recruit new customers with attractive market and resource potential [8, 67].
- *Customer development and retention*: The second objective is to intensify the customer relationships by means of cross and up selling activities, i.e. by proposing the customers other products from the product assortment or higher value goods. Once again, this objective has to be attained in a differentiated way, for instance, unprofitable customers might be redirected to lower value offers (i.e. down selling) which are less costly to handle. On the other hand, customer retention programs have the aim of retaining profitable customers.
- *Customer reacquisition (or recovery)*: The last objective is to reacquire lost customers as, based on the available past information, the reacquisition is potentially easier and less costly as the recruitment of new customers. The recovery process generally focuses on valuable customers. As a result, the company might be pleased if unprofitable customers quit; it has perhaps even wanted so and done some incentives in order to discourage low value customers.

In order to successfully realize these objectives, it is compulsory to have a deep knowledge of the customers and to be able to evaluate the profitability of the customer relationships. The following sections present the concepts which can describe and quantify the customer relationships (Sect. 4.2) as well as the CRM systems and architecture which enable the planning, the analysis and the execution of the market and the customer processing (Sect. 4.3).

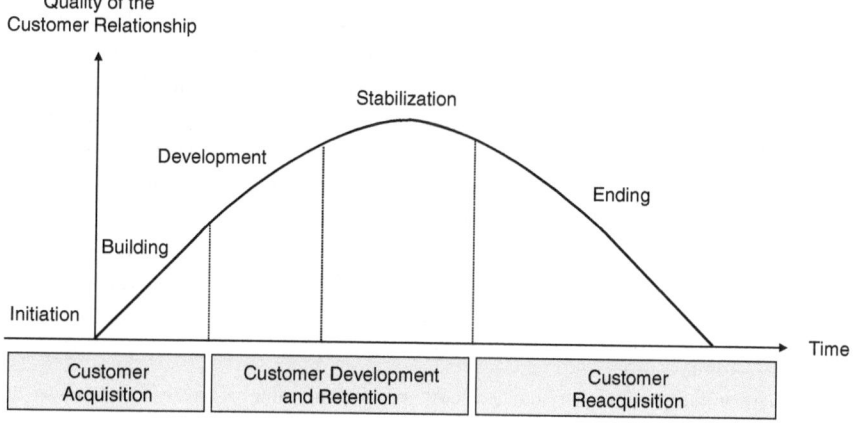

Fig. 4.2 Customer life cycle (adapted from [17])

4.2 CRM Theoretical Constructs

This section presents the main theoretical constructs which are pertinent in this work, some of them have been already encountered in the CRM success chain of Fig. 4.1. Section 4.2.1 looks at the notions of customer value, customer lifetime value and customer equity as Sect. 4.2.2 studies the relationship between the concepts of customer satisfaction, customer loyalty and customer retention.

4.2.1 Customer Value, Lifetime Value and Equity

Since the main objective of CRM is to develop profitable customer relationships, it is essential to be able to measure the value of the customer relationships. Following Hippner [51], the customer value can be perceived either from the customer's perspective, that is the value for the customer called *customer value*, or from the enterprise's perspective in terms of *customer lifetime value*.[1]

Although the customer lifetime value being more relevant to CRM than the customer value, both are intimately linked as shown in Fig. 4.3. By having a customer orientation, companies try to match or exceed the customers' expectations, i.e. to achieve a high customer value (cf. customer satisfaction in Sect. 4.2.2), by proposing adequate products and a high service level. A high customer value is in turn required for building (resp. maintaining) the customer relationships. The higher is the customer value, the more likely it is to have profitable and long term relationships with the customers and consequently a higher turnover, cross selling and recommendation potential, i.e. a higher customer lifetime value. Finally, a high customer lifetime value motivates the enterprises to strengthen the customer orientation and to further optimize the customer related processes.

The customer lifetime value represents the contribution of a customer to a firm. This contribution is not only based on the intensity of the relationship and on the direct contributions but also on the duration of the relationship as well as indirect contributions like the recommendation potential. The customer lifetime value therefore integrates the future potential and also non-monetary assets.

Figure 4.4 shows a hierarchical decomposition of the customer lifetime value proposed by Hippner [51]. This decomposition distinguishes between the transaction potential and the relation potential of a customer. The transaction potential, which represents a customer's actual and future use of a company's services, is derived from the following determinants:

- *Basis volume*: It consists of the buying history of a customer revealing the actual customer relationship intensity. This determinant implies that, based on an already existing customer relationship, the customer has a buying behavior habit as well as

---

[1] In the German literature the here defined 'customer lifetime value' is mostly treated under the name 'Kundenwert'. Since its direct translation refers to 'customer value' which represents in this work the customer perspective, the term 'customer lifetime value' corresponds to the enterprise's perspective.

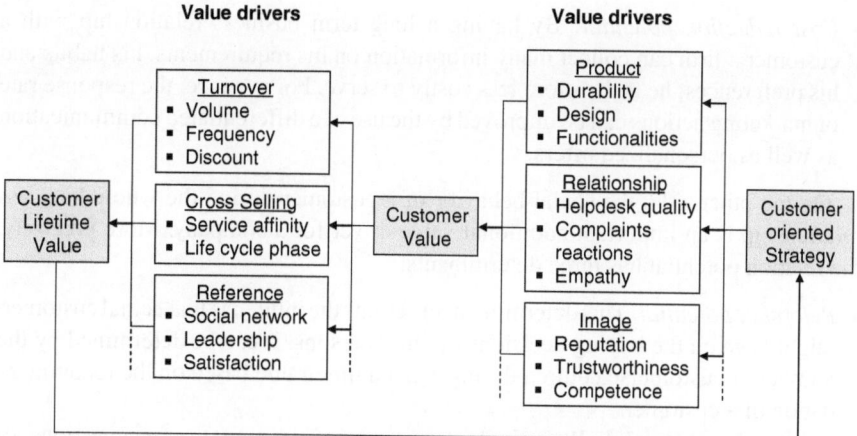

Fig. 4.3 Customer value and customer lifetime value (adapted from [51])

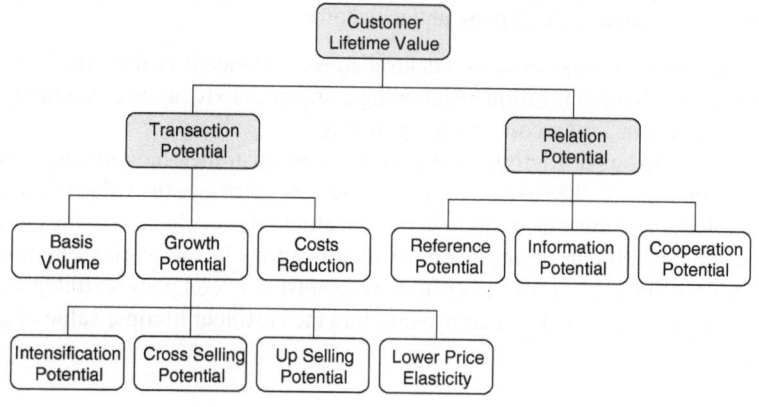

Fig. 4.4 Decomposition of the customer lifetime value (adapted from [51])

an immunization against competitive offers meaning that a relative stabile turnover can be expected in the future.

• *Growth potential*: This determinant represents the expected changes in the buying behavior of a customer, and in turn, can be determined by four determinants. First the intensification potential refers to the future extension or reduction of the basis volume often observed by the repurchase rate. The cross and up selling potentials are the expected additional use of the product offering of the company and the purchase of higher value products respectively. Finally the potential of a decreasing price elasticity with the duration of the customer relationship implies that customers are less sensitive to competitive offers with a short term price advantage or with special conditions.

- *Cost reduction potential*: By having a long term business relationship with a customer a firm can collect many information on his requirements, his habits and his preferences; he is therefore less costly to serve. For instance, the response rate of marketing actions can be improved by the use of a differentiated communication as well as personalized offers.

On the other side, the social behavior of a customer during the whole business relationship is an important additional value driver for a company. More precisely, the relation potential has three determinants:

- *Reference potential*: This determinant represents the influence that actual customers might have on the buying decisions of other persons. It can be determined by the number of customers acquired during a given timeframe based on the recommendation of a customer.
- *Information potential*: By using customers' information, e.g. suggestions or complaints, a firm can optimize its products or services as well as its processes.
- *Cooperation potential*: It consists of the value drivers that are realized through a cooperation between a company and a customer.

The customer lifetime value is a central construct since it enables the long term objective of profit maximization by allowing companies to focus on customer groups having an actual or future contribution potential.

Since the customer lifetime value treats each customer individually, another important indicator is the *customer equity* which encompasses the value of the entire customer base. The customer equity is most useful to detect changes in the customer relationships which can be the result of modifications in the market environment or can reflect the reaction of the customers against new company's strategies. The customer equity can be calculated by summing the customer lifetime value of all the customers.

4.2.2 Customer Satisfaction, Loyalty and Retention

Customer retention is a very important aspect in CRM since the acquisition of new customers is much more expensive than retaining the existing ones [70]. Following Diller [27], the customer retention is defined as repeated transactions of a customer with a given company. The fact that customers buy repeatedly without the company undertaking any acquisition endeavor has three important advantages regarding a company [27]:

- *Security effect*: Customer retention is a security factor since it improves the relationship stability by having customers with a habitual buying behavior, an immunization against other competitors and a greater tolerance to the provider's potential faults. Secondly, the customer retention improves the information feedbacks as bound customers are more willing to express their complaints or their opinions as well as to collaborate actively in the development process.

Customer retention also increases the possibilities and the success chance of interaction means like customer event, customer clubs, etc. Last but not least, the customer retention augments the customer trust towards the company which leads to a consolidation and a deepening of the customer relationship in the future.

- *Growth effect*: Customer retention can also have a turnover growth effect by means of a better customer penetration as well as a development of the customer base. The customer penetration can be achieved by improving the buying concentration, intensity and frequency as well as the cross selling of the customers. On the other side, customer retention can induce a growth effect by developing the customer base through customer recommendations.

- *Rentability effect*: The third aspect of customer retention is an improvement of the rentability due to a cost reduction and to a revenue increase. A costs reduction is possible since bounded customers do no more involve acquisition costs and their handling is less costly. Revenues can also benefit from the customer retention as bounded customers are less price sensitive and, based on the customer knowledge, cross and up selling can be achieved.

According to Hippner [51], the customer retention just as the customer value has two different meanings depending if the customer's perspective or the company's view is taken into account.

From the customer's perspective, the *customer retention* is the aggregation of observable customer behavior patterns and can be modeled as a hierarchical construct proposed by Homburg [57] and shown in Fig. 4.5. The customer retention construct is based on the actual and intended behaviors of a customer. The actual behavior encompasses two determinants, the repurchases and the recommendation of the customers, whereas the intended behavior is based on three determinants, the intended repurchases, the intended cross selling and the intended recommendation.

From the company's perspective, the *customer retention* is a collection of means whose objective is to maintain the customer relationships stable. Homburg and Bruhn [56] proposed a customer retention success chain which integrates the customer retention together with the notions of customer satisfaction and customer loyalty (see Fig. 4.6). The *customer satisfaction* is a central factor for the customer loyalty and, consequently, for the customer retention. It is mostly based on

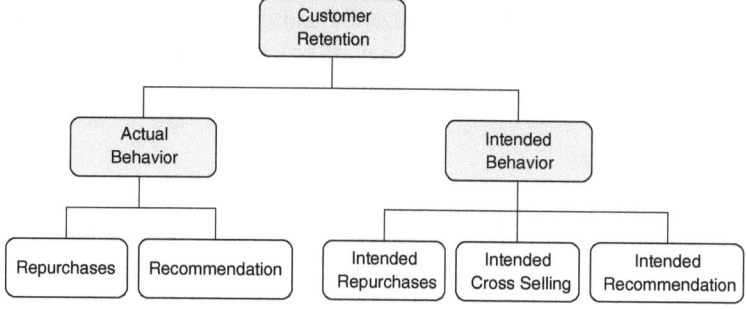

Fig. 4.5 Customer retention construct (adapted from [57])

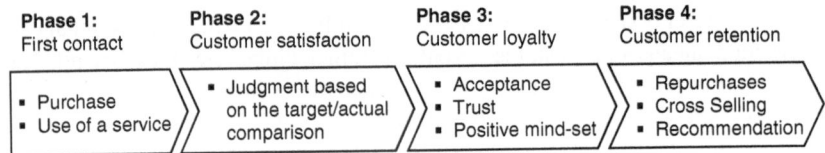

Fig. 4.6 Customer retention success chain (adapted from [56])

the confirmation/disconfirmation paradigm and represents the result of a psychic comparison process between the customer expectations and the effective perceived service level [117]. A customer is therefore satisfied if his subjective expectations are matched or even outmatched (i.e. confirmation) and unsatisfied otherwise (i.e. disconfirmation). Then, the *customer loyalty* depicts a strong orientation of a customer towards a company and therefore influences the customer retention. If a customer has positive experiences with several purchases, he is going to trust the company and will become loyal. Despite the fact that customer satisfaction does not always results into customer loyalty, customer satisfaction is a mandatory prerequisite for customer loyalty. Customer loyalty, in contrast to customer satisfaction, is a long term and stabile construct which can far better predict the future customer behavior [90].

The customer retention success chain shown in Fig. 4.6 has four fundamental phases. This chain starts with a first contact of a customer when buying a product or using a service of a company. Then, in a second phase, the customer is able to compare the product or service against his expectations and draws a satisfaction judgment. If the customer is satisfied, i.e. his expectations are matched or even outmatched, the customer loyalty based on the acceptance, the trust and the mindset of this customer toward the firm emerges in the third phase. Finally, the customer retention can be achieved in the fourth phase when the customer loyalty results in effective repurchases, cross-selling and recommendations.

Following Schaller et al. [101], an effective mean to achieve the above mentioned customer satisfaction, loyalty and retention is the individualization which can be based either on the (core) services of a company, or on the relationship or communication with the customers. Such individual services and communication can be realized by the mass customization which combines the advantages of customization, i.e. individual solutions, and mass production, i.e. low production prices. Mass customization is defined as the customization and the personalization of products and services for individual customers at a mass production price [97].

4.3 CRM Systems

As stated in Definition 4.1, CRM systems are the IT enabler of a customer oriented strategy. By integrating the customer information from the different enterprise's systems, e.g. marketing, sales, and after-sales solutions as well as standard software including enterprise resource planning and supply chain management systems, CRM

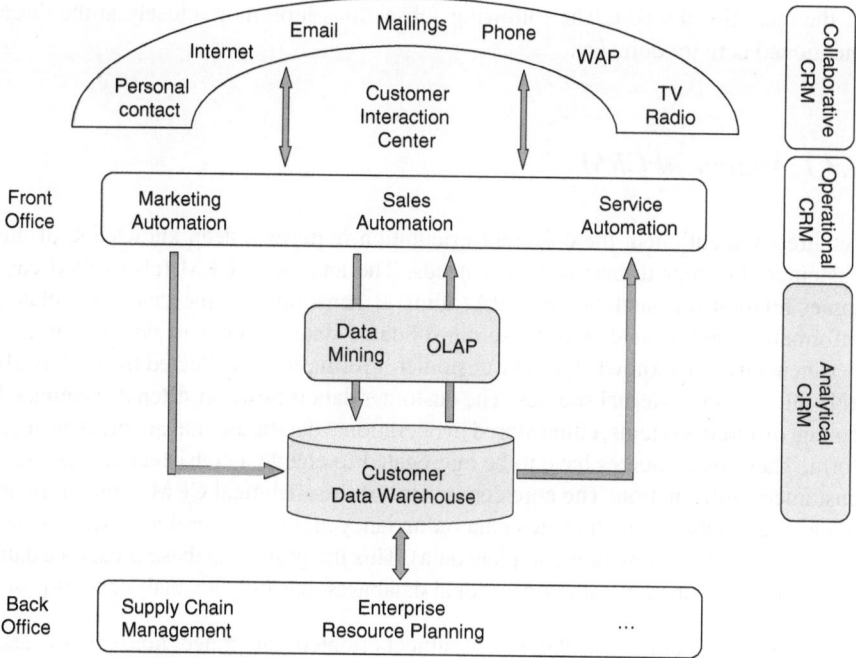

Fig. 4.7 CRM architecture (adapted from [53])

systems enable firstly a global and consistent view of the individual customers (i.e. one face of customer) and secondly a holistic and coherent dialog with the customer base (i.e. one face to customer) [55].

As depicted in Fig. 4.7, the functionalities of CRM systems can be basically divided into three main activities [15, 42]:

- *Analytical CRM*: The role of the analytical CRM is the consolidation and the analysis of the customer related information including the customers' profiles, contact and reaction information. Based on the collected data, its aim is to enable a continuous optimization of the customer related processes. The analytical CRM is based on a closed loop architecture which allows a permanent adjustment of the firm's services and communication to the customers' requirements.
- *Operational CRM*: The operational CRM provides support to the front office processes having a direct contact with the customers, i.e. the marketing, sales and after-sales. By automating these processes, the dialog between the company and its customers can be optimized.
- *Collaborative CRM*: The collaborative CRM is responsible for the integration and the synchronization of all the communication channels in order to allow an efficient bidirectional communication between the enterprise and the customers.

Note that some publications [52, 55] do not distinguish between the operational CRM and the collaborative CRM but consider the collaborative CRM being part of the operational CRM. The following subsections look more closely at the three mentioned activity domains.

4.3.1 Analytical CRM

As already mentioned, the customer orientation requires a deep knowledge of the customers in order to best suit their needs. The analytical CRM (also called customer relationship analytics or CRA) aims at consolidating the customer related information and at analyzing these consolidated data in order to derive new and pertinent customer knowledge. The customer information is collected from all available internal and external sources. The customer data is however often disseminated among different systems, either stored in operational databases or available in printed form. These data sources have to be aggregated to enable a consistent access to all customers' information. The core component of the analytical CRM is therefore an integrated database which avoids data redundancy and related problems (i.e. inconsistent, obsolete, wrong or incomplete data). This integrated database is called a data warehouse and, in contrast to operational databases, has only an analytical purpose.

Definition 4.2 A *data warehouse* is a subject oriented, integrated, non-volatile, and time variant collection of data in support of management's decisions.

In Definition 4.2 Inmon [58] has specified the main characteristics of a data warehouse. A data warehouse aims at modeling specific application's objectives (subject orientation), at integrating data from different databases (integration) which are, once imported, no more modified nor deleted (non-volatility) and at storing historical data (time variance). Following Gawlik et al. [42], different types of customer information can be distinguished in a data warehouse:

- *Basic data*: Contact information, geographic, demographic and psychographic information, buying behavior information as well as buying criteria.
- *Action data*: Channel information, intensity, volume, frequency, dates, firm's activities concerning the customer.
- *Reaction data*: Economic data about the turnover volume and structure or the buying dates as well as non economic data about requests, mindset, knowledge, complaints, duration of the relationship, loyalty level, etc.
- *Potential data*: Product specific global requirement (lifetime value), requirement dates, position in the customer portfolio, customer classification.

In order to support the operational CRM (see Sect. 4.3.2) the analytical CRM, founded on the basic, action and reaction data, can individually assess the customers. In praxis there exist different approaches to evaluate the customers apart the customer lifetime value presented in Sect. 4.2.1, like the qualitative evaluation, the turnover and the contribution calculation, the portfolio analysis as well as the scoring methods [90]. All these approaches produce new customer information which can be stored in the data warehouse (i.e. the potential data). Another essential task of the analytical CRM, based on the customer data (including the potential data), is to segment the customer base into groups of similar customers which can then be addressed in a differentiated way by marketing actions. A customer segment is characterized by a high intern homogeneity and a high external discrepancy [16]. Note that the market segmentation and the customer segmentation have to be distinguished: the market segmentation is basically a partitioning of a whole market while the customer segmentation is limited to the actual customer base.

Following Neckel and Knobloch, two different data analysis approaches exist [90]:

- A *hypothesis driven analysis* (top-down approach) where the objectives are previously known.
- A *data driven analysis* (bottom-up approach) where the expected results are not yet known or clear.

In the case of a top-down approach, the aim of the analysis is to verify or reject the given hypothesis. Different statistical and analytical models can be used to realize a hypothesis driven analysis. Among them, a widely used method following a deductive process, is the online analytical processing (OLAP) which allows the evaluation of indicators (like turnover or sales) in a multi-dimensional data structure. More precisely, it enables an explorative and interactive navigation inside the hierarchies of the predefined dimensions through the use of special aggregation operators (i.e. rotate, drill-down, roll-up, drill-across, slice and dice). Most OLAP tools allow the creation of reports in tabular or graphical form. OLAP tools are therefore very convenient to use and their generated reports are very intuitive to understand.

However, the deductive capability of OLAP limits its use to the analysis of the actual (factual) situation of the company (resp. of the customers). As CRM front end activities often need causal or predictive information in order to optimize their processes, data driven analysis are often required. New knowledge about the customers' data can be derived using data mining techniques following an inductive process.

Definition 4.3 *Data mining* is the process of exploration and analysis, by automatic or semi-automatic means, of large quantities of data in order to discover meaningful patterns and rules [5].

As data mining refers to the actual extraction of patterns and rules using algorithms, the knowledge discovery in databases (KDD) covers the whole process for discovering previously unknown information, including the problem specification, the data preparation, the data analysis, the results analysis and the exploitation of the results [54]. The most often used data mining techniques are decision trees, cluster analysis, artificial neural networks and association rules [90]. These data mining techniques can be used in order to derive valuable information for the different CRM activities. For instance, Hippner and Wilde [55] distinguish different data mining tasks according to the phases of the customer life cycle (see Fig. 4.2):

- *Customer acquisition*: During the customer acquisition phase, even if few information about the potential customers is available, data mining can help optimizing acquisition campaigns. First the *response analysis*, by analyzing already executed campaigns, can determine which customer groups positively reacted. This information can then be used for subsequent campaigns in order to reduce the campaign costs. Secondly, the efficiency of the acquisition campaigns can be improved by a *target selection* which, based on the active customers, can derive customer segments which are very profitable for the company. Marketing actions can afterwards be exclusively focused on potential customers who have a similar profile.

- *Customer development and retention*: Having more customer information available, data mining techniques can be carried out in order to effectively develop and retain the actual customers. The *cross and up selling analysis*, by analyzing the product utilization behavior of previous customers, can propose cross or up selling product propositions for newly acquired and low value customers. In a similar manner, cross and up selling propositions can also be derived using the *basket analysis*. Based on the customers' basket content (i.e. the sets of products bought in single transactions), the basket analysis can determine which product combinations are often sold. More generally, the *customer analysis* allows companies to effectively handle customers based on their profitability.

- *Customer reacquisition*: Apart reacquiring lost customers, a central aspect of the customer reacquisition phase is the detection of likely quitting customers so that concrete measures can be undertaken in order to retain these customers. The *churn analysis*, based on the analysis of already lost customers, can identify the potentially churning customers.

A detailed description on data warehouses, OLAP and data mining is beyond the scope of this work; for additional information please refer to [4–7, 54, 90]

4.3.2 Operational CRM

The operational CRM (also called customer relationship operations or CRO) encompasses all tasks which involve a direct contact with the customers (front office). These tasks are independent of communication channels (see Sect. 4.3.3) and can be divided into three main processes which can be automated leading to the marketing, the sales and the after-sales automations [42]:

- *Marketing automation*: The marketing automation enables the support and the control of the customer related marketing processes. Its aim is to provide a global, logical and predefined management of the customer contacts. The main task of the marketing automation is the campaign management which aims at providing the right information or offer to the right customer with the right communication channel and at the right time. This objective can only be achieved by appropriately using the customer knowledge provided by the analytical CRM. The marketing campaign planning is therefore strongly dependent on the analysis of the customers' characteristics and behavior exposed in Sect. 4.3.1. Once the marketing campaign has been launched, the campaign controlling can assess the customer reactions and reacts with predefined marketing measures. At the end of the campaign, an impact analysis can be done in order to gain some valuable information for future marketing campaigns. A central aspect of the marketing automation is that it not only benefits from the analytical CRM but it also feeds it by storing all the customer reactions as well as the campaign results in the data warehouse, forming this way the closed loop architecture.

- *Sales automation*: A second aspect of the operational CRM is the sales automation which includes the administrative, the analytical and the contact support of the sales tasks. Concerning the administrative aspect, CRM systems can effectively assist employees for the objectives planning, the budgeting, the generation of sales reports, the customer data management and so on. The analytical support of the sales automation can be very useful for enterprises selling on the Internet since the data generated can be utilized for further analysis purposes. For instance, the opportunity management can derive an overview of the actual sales opportunities, the sales cycle analysis tries to predict the date of the customers' next purchase and the lost order analysis, by analyzing all the offers which do not result into a sale, attempts to perceive changes of competitiveness and their origin. Finally, the contact support of the sales automation has the objective of providing the helpdesk (or the customers directly) all the information about the products' features, prices and compatibility.

- *After-sales automation*: The third and last domain of the operational CRM is the after-sales automation. Just like the sales automation, the after-sales automation can be decomposed into the administrative, the analytical and the contact support. Unlike the administrative support which is similar for the sales and after-sales automation, the analytical and contact supports focus centrally on the complaints management. In the case of the analytical support, the number and the form of complaints, the customer satisfaction and the costs of the complaints' processing are important indicators in order to assess the after-sales service quality and to optimize it. The contact support of the after-sales automation is probably the most important component since it has a direct impact on the customers. By the use of a complaints database, in which complaints are stored and treated, employees can quickly find problem solutions or redirect the customers to the appropriate service.

4.3.3 Collaborative CRM

The collaborative CRM (also called customer relationship communications or CRC) has the task of realizing and synchronizing all the communication channels to the customers, e.g. phone, e-mail, Internet, direct contact, and so on. Its objective is to allow an intensive and flexible communication between the firm and the customers so that they have a consistent communication with the company, e.g. customers do not receive duplicated or contradictory information. For this sake, the multi-channel management allows the customers to freely choose and switch between different communication channels during their relationship.

As illustrated in Fig. 4.7, one instrument of the collaborative CRM is the customer interaction center which is an enhancement of the traditional call centers. In contrast to call centers which only handle the phone communication channel, customer inter-action centers also encompass further communication channels like Internet, e-mail, SMS (Short Message Service), etc.

4.4 CRM Systems and Fuzzy Classification

As illustrated in Fig. 4.8, the application scope of the fuzzy classification approach presented in Chap. 3 is clearly located in the analytical CRM (see Sect. 4.3). By providing the operational CRM with a more accurate information on the customers, a fuzzy classification enables a more differentiated handling of the customers compared

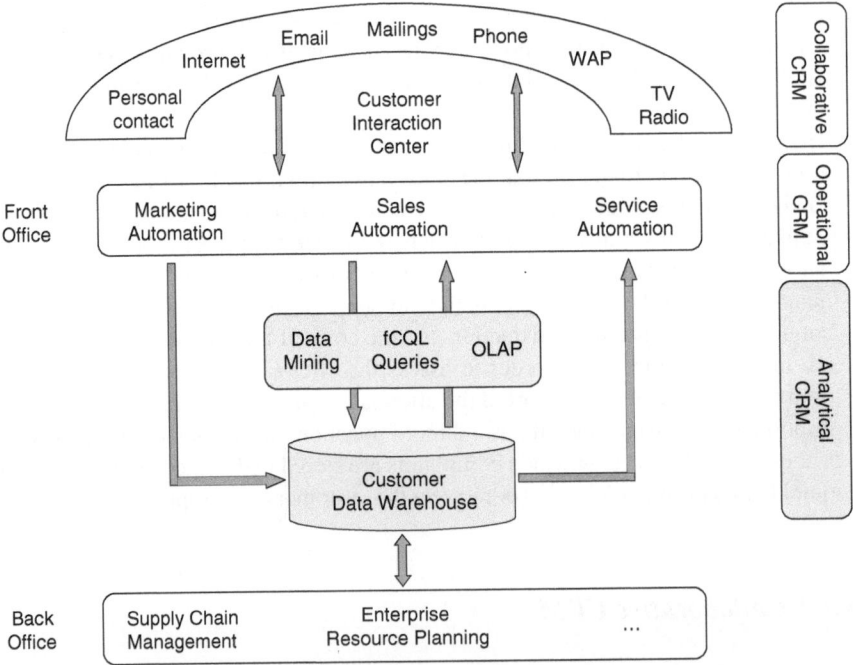

Fig. 4.8 CRM architecture with fuzzy classification querying (adapted from [53])

to traditional classification tools (e.g. market and customer segmentation, portfolio analysis, etc.). The fuzzy classification approach consequently represents a valuable complementary toolkit to the OLAP and data mining techniques.

Different integration approaches can be considered in order to integrate the fuzzy classification information into CRM systems:

- A first approach is to store the classification results and all the related information (see Chap. 5) into the customer data warehouse as part of the potential data (see Sect. 4.3.1). This approach is pertinent if the company wants to classify its customers on a determined frequency, e.g. on a monthly basis. It has the advantages of providing consistent data during the considered period and to be resource effective since the classification is executed only once per period. It also allows the firm to keep the effective classification history of its customers.
- A second integration approach is to dynamically calculate the classification information on a customer base each time this information is required. This method has the advantages of always providing the actual classification information of the customers and to avoid the storage of redundant information since the classification results are derived from customers' data already stored in the data warehouse. Furthermore, if the fuzzy classification definition were to be modified, this approach would automatically reflect the changes for current data as well as historical data.
- A third possibility is to combine the two previous integration approaches in order to benefit from their respective advantages. This method provides a fast access to precalculated historical classification information as well as the flexibility of ad hoc classification queries for more detailed or current information. Of course, this approach also inherits the drawbacks of both previous approaches. For this reason, a decision must be made as to which part of the classification information should be stored in the data warehouse and which part has to be dynamically calculated. The business environment of the firm will finally determine the trade-off between the performance, the flexibility, the data volume and quality requirements.

Chapter 5
Fuzzy Customer Classes

In Chaps. 2 and 3 the potential of fuzzy logic, resp. fuzzy classification, in regards of human beings have been demonstrated. The fuzzy classification approach does not only better model human concepts but it also enables an intuitive human oriented querying process by the means of linguistic variables and terms. Most important, the results of such classification are fuzzy classes where elements can belong to several classes simultaneously and are assigned a precise grade of membership in all the considered classes. As a consequence, the sharp classes' borders disappear, replaced by a continuous transition between the classes. This is a much more real and pertinent projection of the surrounding world, in all-day human activities and especially in business fields where analytical applications are increasingly used.

Chapter 4 then introduced the customer relationship management whose goal, by having a deep knowledge of the customers, is to create, maintain and enhance a long term and profitable relationship with promising and profitable customers. Fuzzy classification is well suited to all management fields which require measurement, assessment and controlling activities, but it is, from the author's perspective, even more attractive in the CRM domain as fuzzy logic is human oriented in nature.

This chapter aims to explicit the benefits of the fuzzy classification approach in the context of the customer relationship management. For this purpose, Sect. 5.1 first demonstrates the weaknesses of the sharp classification approach and motivates the use of fuzzy customer classes. Section 5.2 looks at a direct consequence of the use of fuzzy classes, namely the customer positioning ability. Then, Sect. 5.3 shows how fuzzy classes can automate the mass customization and the personalization. Finally, Sect. 5.4 looks at the hierarchical decomposition, at the customer equity calculation and at the controlling ability of a hierarchical fuzzy classification.

© Springer International Publishing Switzerland 2015
N. Werro, *Fuzzy Classification of Online Customers*, Fuzzy Management Methods,
DOI 10.1007/978-3-319-15970-6_5

5.1 Fuzzy Versus Sharp Customer Classes

Managing customers as an asset requires measuring them and treating them according to their real value [8]. Most analysis methods and tools available on the market, e.g. portfolio analysis, do classify customers sharply. However with sharp classes, i.e. traditional customer segments, it is not possible to assess customers thoroughly [84]. Consider for instance the four customers of Table 5.1. If they were to be sharply classified regarding their turnover and behavior characteristics, customers Smith and Brown would belong to class C_1 whereas customers Ford and Miller would be univocally assigned to class C_4 (see Fig. 5.1).

Using a sharp classification approach implies two shortcomings which conflict with the classification philosophy: on the one hand, customers with similar characteristics can be classified in different classes and, on the other hand, customers with different features can belong to the same class. In Fig. 5.1 for instance, customers Brown and Ford have very similar turnover as well as similar behavior. However, Brown and Ford are treated in totally different classes: Brown belongs to the winner class C_1 (Commit Customer) whereas Ford is assigned to the loser class C_4 (Don't Invest). Conversely, a traditional customer segment strategy treats the top rating customer Smith the same way as Brown who is close to the loser Ford.

Table 5.1 Customer profiles including turnover and behavior

CustomerTable		
Customer	Turnover	Behavior
Smith	1,000	100
Brown	510	51
Ford	490	47
Miller	50	10

Fig. 5.1 Sharp customer classes

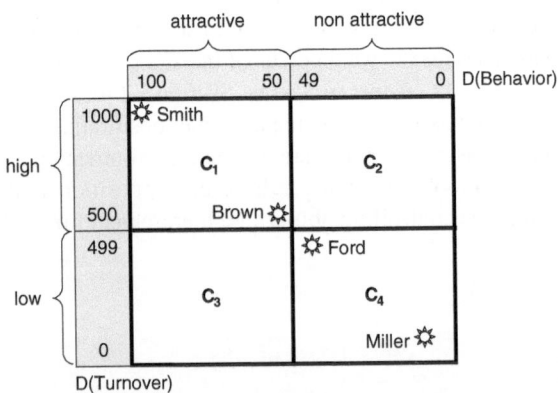

Handling customers sharply, the following drawbacks can be observed:

- Brown, being a medium customer classified in the winner class, has no incentive for improving neither his turnover nor his behavior as he already receives all the privileges of the premium class C_1 (Commit Customer). Furthermore, if his turnover or behavior were to slightly decline, he might be surprised and disappointed as he would fall into another class. He may even fall from the premium class C_1 directly to the loser class C_4 (Don't Invest).
- Customer Ford, who is a potentially good customer, may find more attractive opportunities elsewhere as he is classified in the loser class C_4. He is therefore treated in the same way as customer Miller although he has a higher turnover and a better behavior.
- The most profitable customer with excellent turnover and behavior is Smith. Sooner or later he will become confused. Even though he belongs to the premium class C_1, he is not treated according to his real value. In comparison with Brown, he might be disappointed by the privileges he actually receives.

In order to avoid these problems, one could propose to augment the number of customer classes. This is an erroneous solution since it unnecessarily complicates the customer management and does not solve but simply reduces the inequity problem of the customers between the classes. A sharp classification approach cannot treat customers according to their actual value since sharp classes do not reflect the actual situation of the customers. Sharp classes are consequently discriminatory and should not be utilized for the customers' assessment.

Nevertheless, the mentioned dilemmas can be adequately solved by applying a fuzzy classification as shown in Fig. 5.2. Unlike a traditional classification where customers belong to exactly one class, a fuzzy classification allows customers to

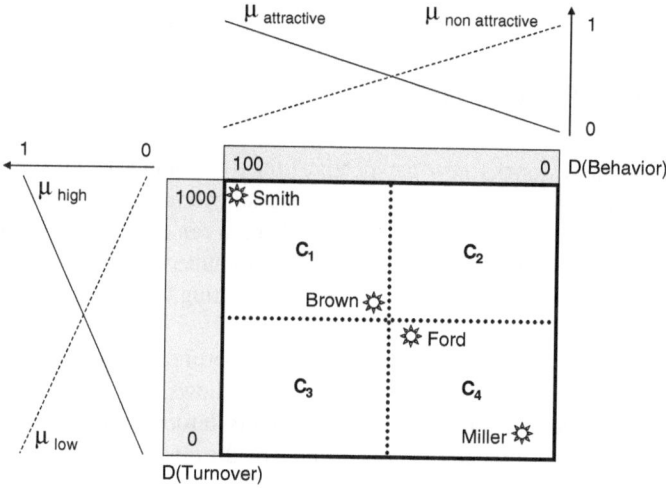

Fig. 5.2 Fuzzy customer classes

belong to several classes at the same time. Furthermore, the customers are assigned a membership degree in every class which indicates to what extent a customer belongs to a given class. A direct consequence is disappearance of the classes' sharp borders and the apparition of continuous transitions between the classes. This way, fuzzy customer classes better reflect the reality and take into consideration the potentials as well as the possible weaknesses of the customers.

Applying the fuzzy classification of Fig. 5.2 (see Sect. 3.2.4), the classification results are:

$$Smith \quad (C_1:1, \ C_2:0, \ C_3:0, \ C_4:0)$$
$$Brown \quad (C_1:0.26, \ C_2:0.25, \ C_3:0.25, \ C_4:0.24)$$
$$Ford \quad (C_1:0.24, \ C_2:0.26, \ C_3:0.24, \ C_4:0.26)$$
$$Miller \quad (C_1:0.02, \ C_2:0.14, \ C_3:0.21, \ C_4:0.64)$$

The results of the fuzzy classification show that customer Smith belongs to only one class as he is standing right in a corner of the classification space. Smith is therefore a core customer of class C_1 and, consequently, the strategy 'Commit Customer' fully applies to him. In contrast all other customers are classified in several classes. Customers Brown and Ford having comparable turnover and behavior are now classified in a similar manner, i.e. their membership degrees in the different classes have a similar value. As they are both located in the middle of the classification space, they are concerned by all the classes' strategies. Note that Smith and Brown are now clearly differentiated and are treated accordingly. Despite the fact that customer Miller is classified in all four classes, his membership degrees indicate a strong belonging to class C_4, therefore the 'Don't invest' strategy is predominant for this customer.

A fuzzy classification is therefore a relevant approach for the evaluation of customers as its results are fair and precise, allowing a company to treat its customers according to their real value.

5.2 Customer Positioning

Having fuzzy classes opens new perspectives for positioning the customers inside the classification space. In contrast to a sharp classification where the only available information is a class belonging, a fuzzy classification can derive the precise position of the customers inside the classes based on their membership degrees. This important information offers new possibilities for segmenting, targeting and controlling customers.

This section aims to depict two concrete and pertinent applications involving the positioning of the customers. Section 5.2.1 first shows to what extent marketing campaigns can benefit from the precise customers information. Then, Sect. 5.2.2 illustrates a new controlling application enabled by the position information, namely the monitoring of customers.

5.2.1 Planning and Optimizing Marketing Campaigns

Marketing campaigns are an effective means for the acquisition of new customers as well as for the win-back of former customers. Marketing campaigns can be, however, very expensive. It is therefore crucial to be able, in a first step, to select the most appropriate customers and, in a second step, to verify the impact of the marketing campaign in order to modify the target group or to improve the strategy of the campaign.

Marketing campaigns are usually based on a customer segmentation where each segment represents a homogenous group of customers, i.e. the customers have similar characteristics, behavior, requirements or taste. An effective segmentation implies that customer segments are internally homogenous and externally heterogeneous [16]. This requirement is unfortunately very difficult to achieve since a segmentation is usually a compromise between the size and the homogeneity of the segments [93]. On the one hand, large customer segments enable economy of scale by having few marketing mix strategies dedicated for a large public but also mean a low internal homogeneity so that the customer targeting may not be efficient anymore. On the other hand, having small customer segments with a strong internal homogeneity comes at a high price because it implies a large number of classes and as many marketing mix strategies.

A fuzzy customer segmentation can overcome the mentioned difficulties by replacing the artificially sharp customer segments by fuzzy ones with continuous transitions. This allows, for instance, a customer to be a potential target for two different marketing mix strategies. Therefore, the fuzzy classification approach offers marketers convenient means for selecting customer subgroups and for measuring the efficiency of the marketing campaign [84].

An example of fuzzy-controlled marketing campaign is given in Fig. 5.3. The chosen strategy is to select customers with attractive behavior and low turnover in order to propose them new or premium products, i.e. cross or up-selling. Using the membership degrees and α-cuts (see Definition 2.9), a subset of customers in class C_3 has been chosen. Once the marketing campaign or a testing process has been launched, the fuzzy customer classes can be analyzed again. It is then possible to verify the impact of the campaign by looking at the position of the targeted customers in the classification space, i.e. if the target group has moved in the direction of class C_1.

Using fuzzy segments, several important implications have to be highlighted:

- Customers are being offered an always larger set of products and services. The differences between products within a given category are becoming difficult to explain to the customers, even for products coming from a single company. It is therefore no more possible to assign each customer a single product or marketing mix as he might be interested in several offers at the same time. Fuzzy customer segments can model such situations and can therefore greatly improve the customer segmentation quality.

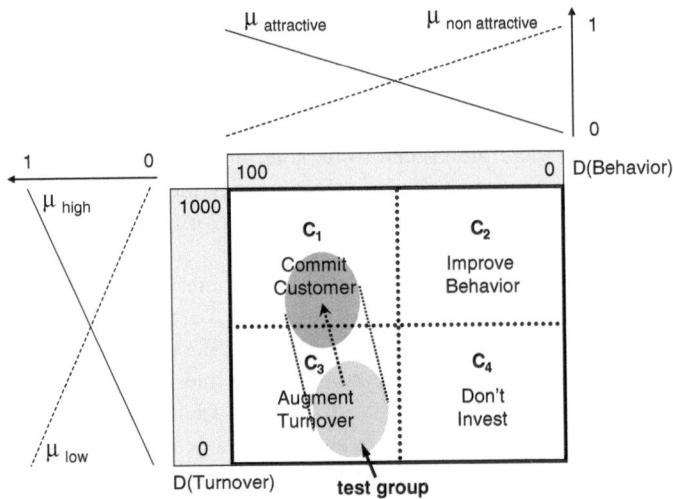

Fig. 5.3 Evolution of the target group due to the marketing campaign

- By having continuous transitions, fuzzy segments overcome the opposition between the size and the homogeneity of the segments. A fuzzy customer segmentation can keep a small number of segments as the homogeneity can be precisely determined with the help of the customers' membership degrees. This information furthermore allows marketers to dynamically modify the size of the target group in order to allow a better planning of the marketing campaigns, for instance to specify the size of the target group with respect to a given campaign budget. Adjusting the size of the target group also enables a more accurate targeting process since it allows marketers to increase or decrease the homogeneity between the targeted customers depending if the proposed products are very specific or intended for a large public.
- Fuzzy customer segments furthermore allow marketers to assess and quantify the impact of a marketing campaign. This controlling can be done during the marketing campaign to verify its evolution over time and, of course, at the end of it to compare the results against the initial objectives. Several measures can be derived from the fuzzy segmentation. The global impact can first be extracted, i.e. how many customers responded and what is the financial consequence. More interesting is the ability of assessing the campaign impact at the customer level, i.e. which customer responded and what was the strength of his response. By detecting customers with a significant reaction, marketers can then select the most promising customers, i.e. a targeting at the customer level. Such an individual targeting is very powerful since the selected customers have proven to be reactive to previous offers. All these measures are very valuable in order to plan and improve the forthcoming marketing campaigns.

5.2.2 Monitoring Customers

Based on the customers' position information, a fuzzy classification allows marketers to monitor the evolution of single customers over time. In contrast to sharp classes where the customers' shifting can only be detected when they move in another class, the fuzzy classes enable the monitoring of the customers' evolution within the classes. By comparing either the attributes' values or the classes' membership degrees of a customer over time, it is possible to detect if this customer is increasing, maintaining or decreasing his value. Marketers can therefore analyze these observations and derive appropriate marketing reactions at the customer level.

A vital and challenging task for companies is the detection of churning customers, i.e. customers who have a likelihood of leaving the company. With the fuzzy classification approach, this task can be automated by implementing a trigger mechanism which supervises the customers' evolution based on several criteria [126]. If for instance a good customer is suspected of having a churning behavior, an alert is sent to the marketing department which can take the appropriate action in order to retain this customer. The retention means to be employed can also be differentiated based on the value and on the churn likelihood of the customer.

Consider for instance customer Williams in Fig. 5.4 who has been acquired at the beginning of 2006 and is, at the end of the first quarter, characterized by a low turnover and a slightly attractive behavior. Due to a cross or up selling campaign, customer Williams improved his turnover over the next three months. Then a loyalty program could enhance his behavior as well as his turnover at the end of September 2006. The development of this customer reached its peak at the end of 2006 due to attractive targeted Christmas offers. Unfortunately, the value of customer Williams dropped in

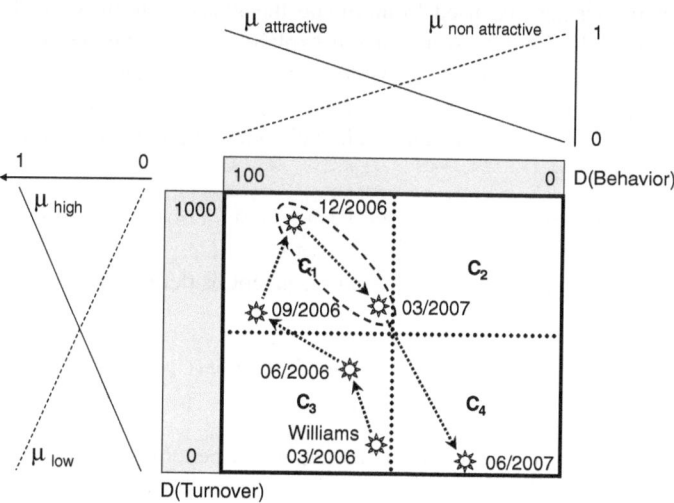

Fig. 5.4 Monitoring single customers over time

terms of turnover and behavior during the first three months of 2007. Finally customer Williams ended his relationship with the company during the second quarter resulting in a turnover near zero. The critical timeframe in this example is not the second quarter when Williams quitted but the first quarter when the first churning indicators could and should have been detected. As at the end of March 2007 customer Williams was still located in class C_1, a traditional classification would have only detected the shift from class C_1 to class C_4 during the second quarter, missing this way the opportunity to efficiently retain customer Williams.

Many other applications can be derived from the customer monitoring which are not restricted to customer retention measures. Such a system could trigger individual offers for cross and up selling and partially replace expensive and sometimes cost-inefficient marketing campaigns. The fuzzy customer monitoring approach also provides a long term perspective of the customers' evolution which allows companies to model the typical customers' life cycle. Afterwards, by knowing the life cycle phases of the actual customers, adequate CRM activities can be scheduled.

5.3 Mass Customization and Personalization

Customization and low cost are often mutually exclusive. Mass production, by producing a large number of identical products, can benefit from economy of scale leading to cost-effective products but at the expense of diversity. Customization, on the other side, achieves personalized offers which generally involve higher prices. Mass customization is defined as the customization and personalization of products and services for individual customers at a mass production price [97].

By providing a precise information about the classified customers, the fuzzy classification approach can be used to automate the mass customization. In a fuzzy classification, each class is given a name explaining its semantics. Based on the semantics of the classes, it is also possible to allocate every class a grade expressing a given concept. The value of a customer regarding this concept can then be achieved by aggregating the grades of the classes he belongs to weighted by his membership degrees in the classes.

Definition 5.1 The *personalized value* $V(O_i|C)$ of an object O_i with membership degrees $M(O_i|C_k)$ in the set of fuzzy classes $C = \{C_1, \ldots, C_n\}$, each fuzzy class having a defined grade $gr(C_k)$ expressing a concept, is defined as

$$V(O_i|C) = \sum_{k=1}^{n} M(O_i|C_k)\, gr(C_k)$$

This section intends to concretely demonstrate the benefits of the mass customization principle in two different cases. First, Sect. 5.3.1 presents a straightforward application of the mass customization by calculating personalized values, for instance a personalized discount. Secondly, Sect. 5.3.2 shows how the mass customization principle can lead to a fuzzification of products (resp. services), i.e. a fuzzy product portfolio.

5.3.1 Personalized Discount

When buying products or services the price is certainly one important decision factor. For this reason companies are using different discount strategies in order to attract and retain their customers. Standard discount methods are based either on the customer (fixed discount per customer sometimes calculated for customer segments), on the quantity (scale price in regard to the quantity of purchased items), on the time (promotional price for a given timeframe), on the region (discount depending on the region) or on other products (special price for a bundle of items) [59].

An enterprise can combine these different methods in order to fit its price strategy. By doing so, the company may encounter the following difficulties:

- Each method has to be defined manually and independently of the other chosen methods.
- The maintenance and adaptation of the discount methods can be difficult since new customers and products are regularly added.
- The combination of several methods may lead to discount collisions which have to be solved.
- With several independent discount methods it is almost impossible to maintain a consistent price strategy on the long run.

To avoid the mentioned problems and to improve the customer retention, a global discount method considering all the aspects of the customers, including the customer's buying attitude and potential, is needed. Such a global discount method can be automatically derived from a fuzzy classification using the mass customization principle, i.e. a personalized discount [125].

Having defined a fuzzy classification, a personalized discount can be easily implemented by assigning discount rates to every fuzzy class. For instance, class C_1 (Commit Customer) has a discount rate of 10 %, C_2 (Improve Behavior) a discount of 5 %, C_3 (Augment Turnover) 3 %, and C_4 (Don't Invest) 0 %. The individual discount of a customer is then calculated using the personalized value definition (see Definition 5.1).

Applying the personalized discount to the four customers discussed in Sect. 5.1 returns the following results:

$Smith\,(C_1:1,\ C_2:0,\ C_3:0,\ C_4:0)$:
$$1 \times 10\,\% + 0 \times 5\,\% + 0 \times 3\,\% + 0 \times 0\,\% = 10\,\%$$
$Brown\,(C_1:0.26,\ C_2:0.25,\ C_3:0.25,\ C_4:0.24)$:
$$0.26 \times 10\,\% + 0.25 \times 5\,\% + 0.25 \times 3\,\% + 0.24 \times 0\,\% = 4.57\,\%$$

Ford $(C_1:0.24,\ C_2:0.26,\ C_3:0.24,\ C_4:0.26)$:

$0.24 \times 10\% + 0.26 \times 5\% + 0.24 \times 3\% + 0.26 \times 0\% = 4.38\%$

Miller $(C_1:0.02,\ C_2:0.14,\ C_3:0.21,\ C_4:0.64)$:

$0.02 \times 10\% + 0.14 \times 5\% + 0.21 \times 3\% + 0.64 \times 0\% = 1.5\%$

Using fuzzy classification for mass customization and personalization leads to a fair and transparent judgment: in contrast to Brown, Smith gets the maximum discount although they are both located in class C_1. Brown and Ford, being very close to each other, receive nearly the same discount rate. The classes' border being fluent, the inequity of the sharp classification disappeared. Finally, Miller benefits from a modest discount rate reflecting his position in the class C_4.

The choice of the personalized discount method over standard discount strategies offers additional advantages to the company apart being easy to manage and fully automated:

- All customers are motivated to improve their buying attitude and/or behavior since they receive a higher discount rate. The customers can also concretely see their progression and incentives like "if you buy this product you'll get a discount of ..." can be used.
- Considering that many companies generate a large part of their profits with only a relative small fraction of their customer base [51], i.e. only a small group of customers are located in class C_1 and even less fully belong to it and get the maximum discount, the total budget for personalized discounts is generally much smaller compared with conventional discount methods.

In the case of our four customers, being almost homogeneously distributed (i.e. an unlikely case), the personalized discount has a cost of 145.52 Euro ($1000 * 10\% + 510 * 4.57\% + 490 * 4.38\% + 50 * 1.5\%$) compared to the cost of 151.− Euro ($1000 * 10\% + 510 * 10\% + 490 * 0\% + 50 * 0\%$) for sharp classes having the same discounts. The actual cost of the personalized discount strongly depends on the distribution of the customers in the classification space and on the discount rates assigned to the classes. The savings can then be used either for augmenting the discount rates of the classes, i.e. to reinforce the customer retention and add-on selling, or can be allocated for acquisition purposes (see Sect. 5.2.1).

The personalized discount approach is a powerful means for customer retention as the retention incentives, i.e. the discount rates, are always proportional to the attractiveness of the customers towards the company. Furthermore, personalized discounts push add-on selling by motivating the customers to improve themselves in order to receive more privileges. Note that it is possible to combine the personalized discounts with methods based on the time or on bundle of products in order to acquire new customers.

The calculation of a personalized discount is an obvious and pertinent example but the mass customization principle is not limited to it. Many other concepts can be derived depending on the company's activity like a risk level, an interest rate as well as own funds for credit worthiness, a customers' priority which can be used for instance for managing the queue of a call center and, more generally, higher level concepts for decomposition purpose (see Sect. 5.4.1).

5.3.2 Fuzzy Product Portfolio

A logical extension of the mass customization principle is the customization of the products (resp. services) themselves [80]. This means that customers are not choosing a classical product with pre-defined characteristics anymore but a fuzzy product whose specifications are automatically adapted to the customer's taste and/or requirements. The fuzzification of products has become possible thanks to the digital revolution which allows digital products to be dynamically generated at (almost) no cost [59].

This subsection aims to explicit a concrete example of a fuzzy product portfolio from the telecom industry. Many telecom companies propose, besides the classical model where the customers pay for the communication time plus a fixed tax, a subscription model including free conversation time and SMS (Short Message Service). These subscriptions aim to attract and retain the customers with attractive prices compared to the traditional cost models and by the fact that these subscriptions contractually bind the customers for a given period. These offers are also appealing for companies: firstly, they can precisely forecast the minimal earning of the subscribed customers; secondly, the profitability of these offers can even be higher than the traditional model when customers do not match the offers, i.e. if the customers exceed or do not use the included conversation time (resp. the included amount of SMS).

Figure 5.5 shows a classical telecom product portfolio. This portfolio distinguishes between the included conversation time ranging from 0 to 120 min and the included

Fig. 5.5 Classical telecom product portfolio

	Included SMS		
Included conversation time	120 minutes & P$_1$ 200 SMS	120 minutes & P$_2$ 100 SMS	120 minutes & P$_3$ 0 SMS
	60 minutes & P$_4$ 200 SMS	60 minutes & P$_5$ 100 SMS	60 minutes & P$_6$ 0 SMS
	0 minutes & P$_7$ 200 SMS	0 minutes & P$_8$ 100 SMS	0 minutes & P$_9$ 0 SMS

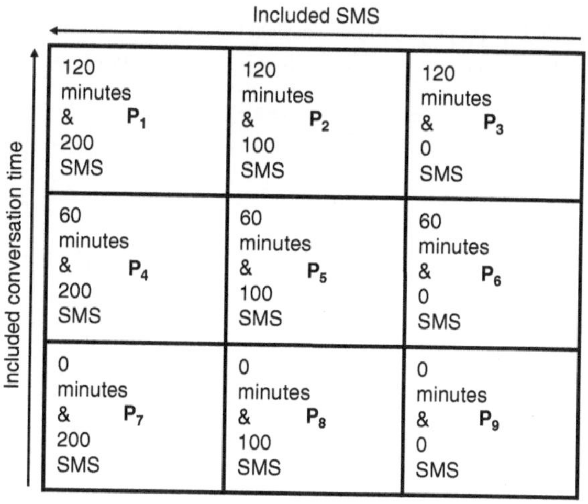

number of SMS going from 0 to 200. The subscription P_1 is the most extensive offer with 120 min and 200 SMS included. At the other extreme, the subscription P_9 represents the traditional model without any extra. The monthly fee as well as the prices for additional conversation minutes and SMS are given for each subscription in Table 5.2.

The subscription model is however only interesting for a customer if he can accurately determine an average conversation time (resp. an average number of SMS) per month. In order not to pay an excessive price, this average should also not vary much from month to month. Since customers who have chosen the wrong subscription might be very dissatisfied, telecom companies started so-called 'fairpay' programs which inform their customers about the best suited subscription based on their call behavior. This is unfortunately a second hand solution since customers with an average conversation time of 90 minutes per month have only the choice between the 60 or 120 min subscriptions. In the first case the user pays at a high cost the extra 30 min and in the second case he does not use the whole conversation credit.

An elegant solution would be to introduce a fuzzy product portfolio which does not suffer from sharp class borders anymore. With the fuzzy product portfolio, the customers are always offered the subscription they actually need. Two different methods can be used to position the customers within the portfolio:

- In the first method, the conversation time and the SMS of the current month are used, in which case the customers always match the personalized subscription. This is the ideal case from the customer perspective but may not be the most profitable one for the telecom companies since no additional costs can be added to the subscription fees. The problem could be solved by fixing higher prices for all the subscriptions.
- The second method calculates the average conversation time and the average number of SMS of the customers on a given period (e.g. over 12 or 24 months) so that the personalized subscription follows the customers' behavior over time. In this

Table 5.2 Properties of the classical telecom subscriptions

Subscription	Monthly fee	Price per additional minute	Price per additional SMS
P_1	40.–	0.30	0.15
P_2	35.–	0.30	0.15
P_3	30.–	0.30	0.10
P_4	30.–	0.30	0.15
P_5	25.–	0.30	0.15
P_6	20.–	0.30	0.10
P_7	20.–	0.20	0.15
P_8	15.–	0.20	0.15
P_9	10.–	0.20	0.10

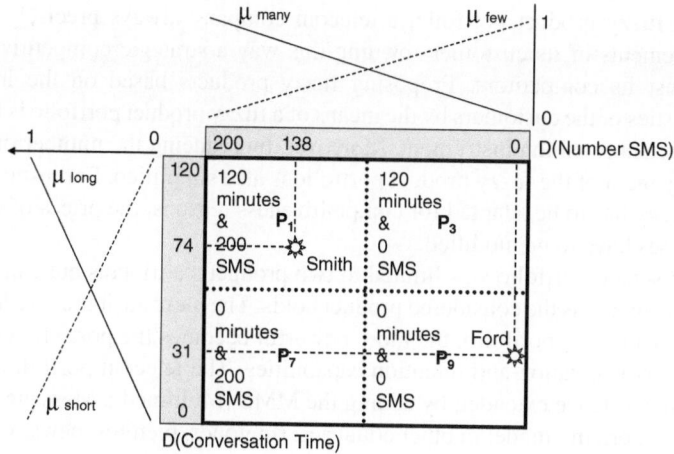

Fig. 5.6 Fuzzy product portfolio

case, the customers still have to pay extra costs if they exceed the offer's limits. The second method is considered further in this example.

Figure 5.6 shows the fuzzy product portfolio which is derived from the classical one. As there is a continuous transition between the classes, the intermediary classes of Fig. 5.5, i.e. P_2, P_4, P_5, P_6 and P_8, can be dismissed simplifying the product portfolio. The four resulting classes represent fuzzy products, meaning that their characteristics can be combined in order to match the customers' requirements.

In Fig. 5.6 the personalized subscriptions of customers Smith and Ford are illustrated. Smith, who is an active customer, has an average conversation time of 74 min and an average number of 138 SMS per month. In contrast, customer Ford is a moderate user who only phones with an average conversation time of 31 minutes per month. The fee of the subscription and the prices for the additional minutes and SMS can then be calculated following the mass customization principle. In this example, the fees of the personalized subscriptions for Smith and Ford are respectively 27.82 and 17.42 Euro per month, the additional minutes cost 0.26 and 0.24 Euro and the additional SMS cost 0.13 and 0.10 Euro. The belonging of Smith and Ford to the fuzzy portfolio classes as well as their personalized subscriptions are summarized in Table 5.3.

Table 5.3 Personalized subscriptions using a fuzzy product portfolio

Customer	P_1	P_3	P_7	P_9	Monthly fee	Minute	SMS
Smith	0.36	0.22	0.27	0.15	27.82	0.26	0.13
Ford	0	0.37	0	0.63	17.42	0.24	0.10

With a fuzzy product portfolio, a telecom company always precisely matches the requirements of its customers owning this way a strategic competitive advantage against its competitors. Proposing fuzzy products based on the individual characteristics of the customers by the means of a fuzzy product portfolio is therefore a very efficient retention instrument. Moreover, by reducing the number of classes, the management of the fuzzy product portfolio is also simplified. For instance, if the price strategy has to be adapted for competitiveness reasons, the prices of only four fuzzy classes have to be modified.

The presented portfolio was limited to two product's attributes but can integrate as many attributes as the considered product holds. The more attributes are integrated in the fuzzy product portfolio, the more powerful becomes the portfolio concept in terms of personalization and retention capabilities. The telecom portfolio example could be for instance extended by adding the MMS (Multimedia Message Service), data transfer, pricing model in other countries, ringtones, pictures, news, voice messages, etc.

5.4 Customer Assessment and Controlling

Most of the concepts presented in this chapter supposed that customers could be evaluated based on their turnover and their behavior. This is of course for presentation purpose only. Indeed, a large number of indicators are available to precisely characterize the customers. Zumstein for instance proposed a CRM success chain involving more than 170 key and classical customer performance indicators [141]. The choice of these indicators is very important as it determines the scope and the capabilities of the fuzzy classification. For a company, many factors guide this choice like the industrial field, the enterprise's culture, the availability of the customer related data, the number of customers and employees, the IT knowledge and budget as well as the desired outcome.

This section aims to give a more realistic view of the implementation of the fuzzy classification approach on a logical level, leaving the detailed analysis and modeling of online customers with a concrete example for Chap. 6. For now, Sect. 5.4.1 first explains how complex classifications can be decomposed into a hierarchy of fuzzy classifications in order to build more consistent and valuable concepts. Based on the hierarchical decomposition, Sect. 5.4.2 shows how a company can, on a customer level, derive the customer lifetime value and, on the company level, calculate the customer equity. Then, Sect. 5.4.3 concludes this chapter by showing how the controlling of customer relationships can be achieved using the fuzzy classification approach.

5.4.1 Hierarchical Decomposition

Customers can be classified considering many different attributes. The classification example shown in Fig. 5.2, where customers are only assessed regarding turnover and behavior cannot be effective in reality. On the other hand, a fuzzy classification with many linguistic variables and terms leads to a multi-dimensional classification space with a large number of classes. By combining all the available attributes, it may not be possible to extract a clear semantics for the resulting classes. This problem, also true for sharp classifications, is partially resolved by the use of fuzzy classes, i.e. by having a continuous transition between the classes, fewer equivalence classes (linguistic terms) are required, reducing this way the number of final classes. The complexity problem remains however for the number of dimensions (linguistic variables) which exponentially increases the number of classes.

In order to maintain a small number of classes with a proper semantics, it is possible to decompose a multi-dimensional fuzzy classification into a hierarchy of fuzzy classifications [126]. By regrouping attributes of a given context into sub-classifications it is possible to derive a precise definition of the classes and, at the same time, to build new concepts expressing a higher semantics than the basic attributes taken separately. It also reduces the complexity of the initial problem allowing a better definition and optimization during the modeling phase. A good example of decomposition and a central perspective for a company is the customer loyalty. Many loyalty concepts have been proposed in the marketing literature. Harrison, for instance, proposes to express the customer loyalty based on two dimensions, the attitude and the behavior of the customers as shown in Fig. 5.7 [48].

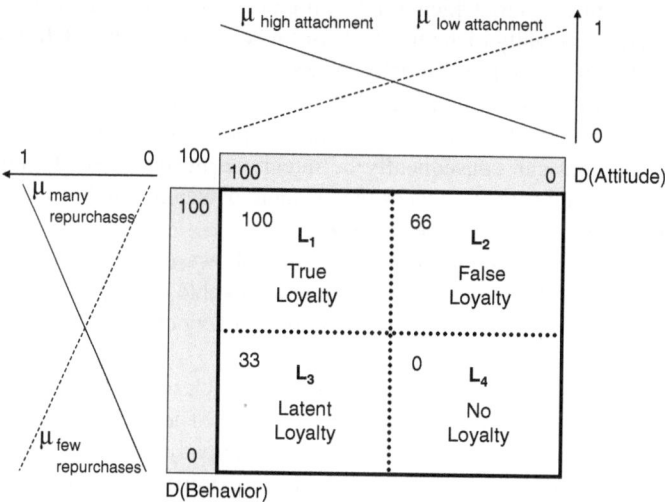

Fig. 5.7 Fuzzy classification for the concept of customer loyalty

The attitude dimension expresses the relative attachment a customer feels towards a company whereas the behavior dimension indicates the customer's repurchase rate. The four resulting classes L_1 to L_4 express different degrees of loyalty [48]:

- *True loyalty*: The true loyalty is characterized by a high attachment and a high level of repurchases. Such customers are very valuable since they are proud to be associated with the company and are pleased to share their satisfaction with others becoming therefore advocates of the firm and its products and services. These customers are also less sensitive to competitive offers and special promotions.
- *False loyalty*: Customers showing a high repurchase rate and, at the same time, not feeling a high attachment to the firm can be characterized as having a false loyalty. Such customers seem to be totally loyal from a purely behavioral perspective even though they might not be at all. Several factors, like a weak market competition, high switching costs, proprietary technologies as well as a customers' inertia or indifference can explain that unsatisfied customers are not willing to change for another competitor.
- *Latent loyalty*: Latent loyalty is also paradoxical in the sense that customers having a high attachment to a firm do not concretize it by repeated purchases of the firm's offering. Situational factors like distribution or convenience problems can inhibit the purchasing behavior. Customers having a latent loyalty might also be patronizing other suppliers.
- *No loyalty*: Finally, customers showing a low attachment as well as a low repurchases rate have no loyalty towards the company. The low attachment of the customers can be due to an inefficient company's communication or to the market environment.

The combination of the attributes attitude and behavior leads therefore to the concept of customer loyalty which is a very important indicator for an enterprise. By assigning grades of loyalty to the fuzzy classes it is possible, using Definition 5.1, to derive the individual customers' loyalty towards the company. The loyalty grade is a new customer information which is not derived from the customers' data directly but is calculated by means of the mass customization principle presented in Sect. 5.3.1. The loyalty concept can consequently be integrated in the fuzzy classification of Fig. 5.2 as it advantageously replaces the attribute behavior which was too weak to express the fidelity of the customers (see Fig. 5.8).

In other words, the decomposition principle allows marketers to merge given attributes in order to build more consistent and valuable concepts. These concepts can then be integrated as new dimensions in other fuzzy classifications leading to a hierarchy of fuzzy classifications.

The decomposition of a multi-dimensional fuzzy classification can be achieved following a 'top-down' or a 'bottom-up' approach. The 'top-down' approach is used when the semantics of the final classes are already known. In this case, the dimensions which define the classes have to be determined. The dimensions can be concrete attributes or new concepts provided by sub-classifications. This process is repeated until all the fuzzy classifications are based on concrete attributes. The 'top-down' approach is particularly useful if the fuzzy classes have to be mapped with

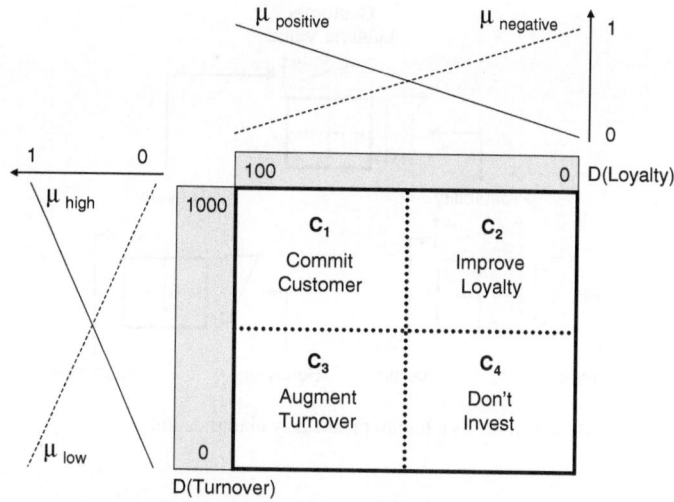

Fig. 5.8 Replacement of the attribute behavior by the loyalty concept

the terminology of the marketing department. The 'bottom-up' approach is more pragmatic by analyzing the available attributes and by combining them to create new concepts until the top fuzzy classification is reached. This approach is meaningful if the top classification is not yet defined. In real cases, both approaches are often used together due to the fact that the final classes defined with the 'top-down' approach sometimes require attributes which are not available.

5.4.2 Customer Lifetime Value and Equity Calculation

In order to manage customers as a company's asset, it is necessary to measure, to treat and to maximize them according to their real value [8, 100]. For this reason companies need means to determine the real value of their customers, i.e. the customer lifetime value. As discussed in Sect. 4.2.1, the customer lifetime value encompasses all the customers' contributions to the firm, including the actual and future potential as well as the direct and indirect contributions of the customers. These contributions can furthermore be monetary, i.e. turnover, margins, profitability, etc., or can be intangible assets, i.e. attachment and loyalty. These different contributions can then be aggregated in order to determine the customer lifetime value (see for instance the decomposition proposed by Hippner in Fig. 4.3). Such an aggregation can be implemented using a hierarchical fuzzy classification where the customers' contributions can be used to derive higher value concepts (e.g. customer loyalty) up to the top fuzzy classification representing the customer life time value (see previous section).

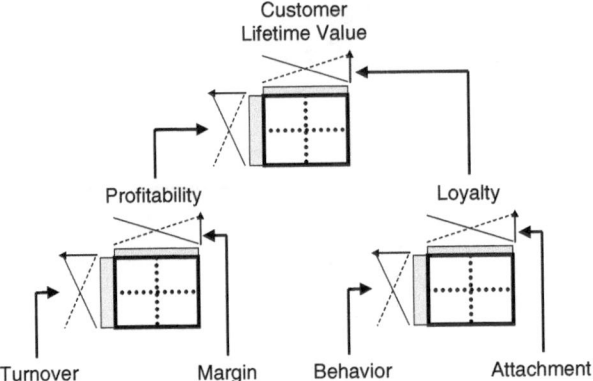

Fig. 5.9 Customer lifetime value as a hierarchical fuzzy classification

For illustration purpose, consider the hierarchy of fuzzy classifications depicted in Fig. 5.9. In this example, the customers are evaluated based on four available attributes, i.e. the turnover, the margin, the behavior and the attachment. Since the combination of these four attributes in one fuzzy classification would not lead to meaningful classes, the attributes have been grouped based on their respective contexts. As explained in the previous section, the attributes behavior and attachment can be combined in order to derive the concept of customer loyalty. In a similar way but in a financial context, the attributes turnover and margin lead to the concept of profitability. Finally the new available loyalty and profitability information can be combined in the top fuzzy classification in order to calculate the lifetime value of the customers.

Thus, based on the available customer information, a customer lifetime value construct and the hierarchical decomposition principle, companies can automatically calculate the individual lifetime value of their customers. The customer equity being the value of the entire customer base, it can be calculated by summing the lifetime values of all the customers.

5.4.3 Controlling of the Customer Relationships

As previously described in the present chapter, the customer positioning and the mass customization are powerful means for the acquisition, the development and the retention of customers. However, companies also need means for controlling and evaluating CRM activities. Following Meier [76], not only the company's management but also the stakeholders are interested on information about the development of the firm's value as well as the utilization of the enterprise's resources. For instance, the following questions can be raised [76]: 'How does the customer base evolve?',

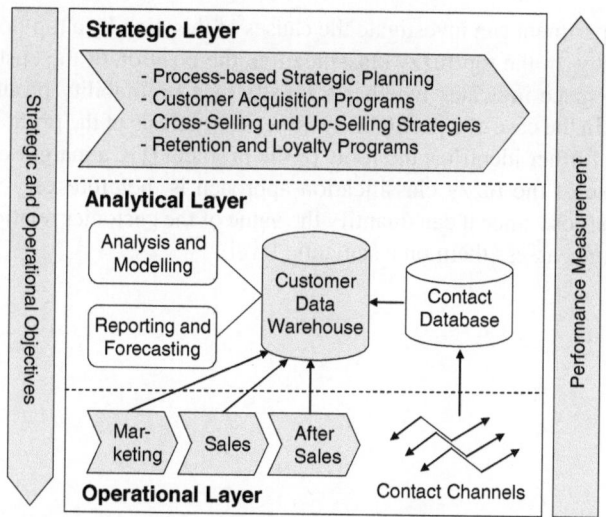

Fig. 5.10 Controlling loop for customer relationship management [81]

'Is the customers' potential exploited?' or 'Does the customers lifetime value progress in relation to the undertaken CRM strategies?'.

Figure 5.10 illustrates the closed loop for controlling customer relationships. At the strategic level, objectives for customer acquisition, retention, and add-on selling must be defined, as must also the process and service quality goals. The traditional tasks in marketing, sales and after-sales activities are carried out in the operational layer. In addition, all customer contacts information have to be analyzed and stored in a contact database. The glue between the strategic and operational layers is the customer data warehouse extended by the contact database. It is also in the analytical layer that the analysis of the contacts and the behavior of customers takes place.

The previous subsections presented how customers can be assessed using a hierarchical fuzzy classification. The management is therefore able to compare the evolution of different customer indicators over time, either individually or on the whole customer base, and consequently, to precisely analyze the impact of the CRM activities. It is important to note that not only the customer lifetime value can be evaluated but all the concepts used in the hierarchy of fuzzy classifications can also be addressed separately. In the example of Fig. 5.9, the customer lifetime value as well as the customers' profitability and loyalty can be assessed.

The use of a hierarchical fuzzy classification for the control of the customer relationships allows managers not only to extract the precise value of the customers but also to analyze the potential and the weaknesses of the classified customers in the different levels of the hierarchy. As each level of the hierarchy expresses a concept defined by a fuzzy classification (i.e. each class having a clear semantic definition), it is possible, for each customer, to derive the appropriate measures in order to improve his value. If, for instance, the lifetime value of a given customer suddenly drops, the marketing department can investigate the causes of this drop by a top-down analysis of the hierarchy. In the top fuzzy classification, the position of the customer in the classification space indicates whether a loyalty or a profitability problem (maybe both) occurs. In the case of a profitability issue, the analysis of the profitability fuzzy classification further identifies the roots of the problem (i.e. a margin or a turnover drop). The use of the fuzzy classification approach is therefore very valuable for controlling purpose since it can quantify the value of the customer relationships and, at the same time, assess them on a semantic level.

Part III
Application and Implementation Perspective

Chapter 6
Fuzzy Classification Applied to Online Shops

Since the previous chapters presented the fuzzy classification approach only on a conceptual level, this chapter aims at demonstrating a concrete implementation. This implementation takes place in the e-business application field as online shops, and consequently online customers, are considered. The choice of online customers is primarily due to the fact that many information about the actions and the behavior of such customers are generally available. This information can furthermore be extended with data coming from back office systems like enterprise resource planning and supply chain management systems.

In order to get a representative case study, a partnership with a small-sized wholesaler company located in Bremen (Germany) has been created. The Kiel & Co company is selling protective gear to business and private customers. Kiel & Co had a website presenting their product catalog online but since it did not have an online shop solution, no online transactions were possible. All the orders were therefore passed by phone, fax, mail and e-mail. In 2005 Kiel & Co adopted the eSarine webshop developed in the Information Systems research group of the University of Fribourg [37, 124, 127] in order to enable the complementary electronic selling channel. Even though most of the transactions still occur on the traditional channels, enough data could be collected in order to model a hierarchical fuzzy classification of their online customers.

This chapter is structured as follows: Sect. 6.1 gives first an overview of the e-business framework. Section 6.2 then refers more specifically to the electronic commerce by considering online shops for small and medium-sized enterprises. Section 6.3, based on the data collected from Kiel & Co, discusses and motivates the chosen hierarchical fuzzy classification. Finally Sect. 6.4 precisely analyzes the case of four typical Kiel & Co customers in all the levels of the hierarchy of fuzzy classifications.

© Springer International Publishing Switzerland 2015
N. Werro, *Fuzzy Classification of Online Customers*, Fuzzy Management Methods,
DOI 10.1007/978-3-319-15970-6_6

6.1 The E-Business Framework

The electronic business generally refers to the use of information and communication technologies (ICT) for the support of business transactions. Many definitions of e-business can be found in the literature, all include the use of ICT for business transactions but some of them also include the notion of actors and added value for the business transactions. Since e-business is often used as a generic term encompassing all types of business transactions, Fig. 6.1 clarifies the situation by distinguishing between the possible actors, i.e. the company itself, the business partners and the customers. This distinction allows e-business to be differentiated into the notions of e-procurement and e-commerce, depending on whether the company is dealing with business partners in order to purchase goods or services, or with customers to sell products or services.

In the e-business framework, the electronic business encompasses all the business transactions from the enterprise's perspective, including the e-commerce and the e-procurement. Following Schubert [104], the e-business focuses on:

- *Information and communication technologies*: The Internet (resp. the intranet and the extranet) can be used to transmit information world wide in an efficient and cost effective manner.
- *Relationships*: The use of information and communication technologies significantly improves the communication and the transactions as well as the exchange of documents between employees, business partners and customers.
- *Integration*: The realization of (electronic) interfaces between legacy systems and Internet based e-business applications allows business partners to be integrated in the business processes.
- *Collaboration*: The collaboration with business partners can lead to a leverage effect. For instance, complementary goods offered in an e-commerce application can extend the firm's products or services catalog in order to better serve its customers.

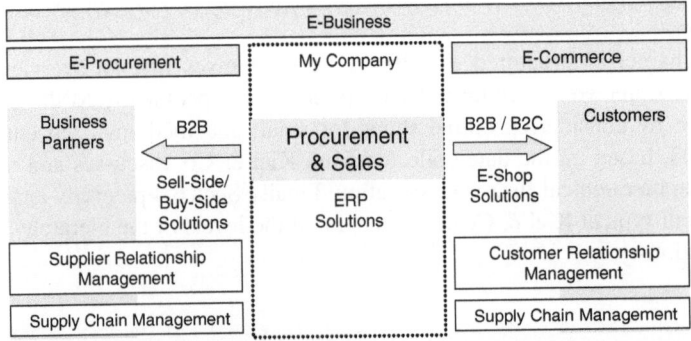

Fig. 6.1 E-business framework (adapted from [107])

As already mentioned, e-commerce is related to the selling process of a company. The customers can be either private persons, leading to a business-to-customer (B2C) relationship, or business partners, leading to a business-to-business (B2B) relationship. A customer to customer (C2C) relationship is also possible if a private person is considered instead of a firm. Generally the e-commerce is driven by the customer relationship management (see Chap. 4) and is supported by online shops (see next section).

As part of the e-business, e-procurement concentrates on the relationship with suppliers for the acquisition of products or services. In contrast to e-commerce, e-procurement involves generally exclusively business partners; it is consequently characterized by a B2B relationship. E-procurement normally relies on managerial methods like the supplier relationship management and the supply chain management.

Since the e-business framework of Fig. 6.1 is a simplification of the original framework proposed by Schubert et al. more information can be found in [104, 107–109].

6.2 E-Commerce and Online Shops

As introduced in the previous section, e-commerce refers to the sales of products and services by means of information and communication technologies. In that context, electronic markets offer new perspectives which can be characterized by the following properties [75]:

- *Place and time independence*: Electronic transactions can be passed from everywhere in the world and at anytime as long as an internet access is provided.
- *Market transparency*: Since everyone has a global access to the information about the products and services, the market's players are forced to eliminate or to justify regional price differences.
- *Reduction of transactions costs*: During the information and the agreement phases important savings can be achieved since these phases can be fully automated. In the settlement phase, costs reduction can also apply if digital goods are considered.
- *Interactivity*: The electronic markets support a bidirectional communication between the producer and the consumer by means of a combination of texts, images, sounds, animations, etc.
- *Responsiveness*: The new definition of the value chain in electronic markets allows more competent and efficient processes. The use of new technologies, like digital agents, for the support of customers can also lead to an improved responsiveness.

In order to benefit from the above-mentioned new electronic markets properties, online shops can be used. Since the subject of the case study, the Kiel & Co company, is a small-sized enterprise, Sect. 6.2.1 illustrates the pertinence of electronic shops for small and medium-sized enterprises (SMEs). Then, Sect. 6.2.2 depicts the common processes and repositories of online shops.

6.2.1 Online Shops for SMEs

In recent years electronic commerce has become increasingly popular since customers are offered a new and easier way of getting the desired information, of comparing prices and functionalities of products and services as well as convenient payment and delivery methods. E-commerce becomes therefore an important competitive advantage for the SMEs.

From the European e-Business Report 2006/2007 [32] some illustrative figures can be extracted: 98 % (resp. 99 %) of the small (resp. medium) sized enterprises have Internet access; this proportion comes to 75 % (resp. 83 %) if the broadband Internet access is considered. More interesting is that 26 % (resp. 29 %) of small (resp. medium) sized enterprises accept orders from online customers. However the percentage of small (resp. medium) sized enterprises using ICT solutions for selling online falls to 12 % (resp. 16 %). As these statistics show, it is actually ever more important to be accessible online because we are moving from a traditional society to an information oriented one. In contrast to what has been thought in the 90s, electronic commerce will not replace the traditional commerce but will become a vital complement of it.

When products are to be presented and sold on the Internet, electronic shops offer a cost effective solution for SMEs. Customers visiting a so-called electronic storefront can choose the products they are interested in, put them in a virtual cart and order those products online [119]. Electronic shops also enable new ways of retrieving precisely information about offered products. Features like product search by keyword, detailed description of a selected product, product display by category and product pictures display are implemented in almost all online shops. Advanced features, like the adaptation of the storefront appearance [73], the recommendation of products [71, 106, 113, 114] or mass customization means, i.e. a personalized discount (see Sect. 5.3.1), furthermore allow SMEs to build and maintain a virtual relationship with their customers which can compensate, at least partially, the personal contact with the customer base. All these functionalities combined with the properties of the electronic markets mentioned in the previous section are undeniable advantages over the traditional remote selling channels like the telephone or the fax.

6.2.2 Webshop Processes and Repositories

A webshop is a web-based software system that offers goods and services, generates offers, accepts orders with different modes of payment, and makes delivery. In principle, each webshop consists of a storefront and a backfront. The online customers only have access to the storefront and can seek information on products and services, place orders, make payment and receive their product. In contrast to the storefront, the access of the backfront is reserved to the webshop administrator. Here, products

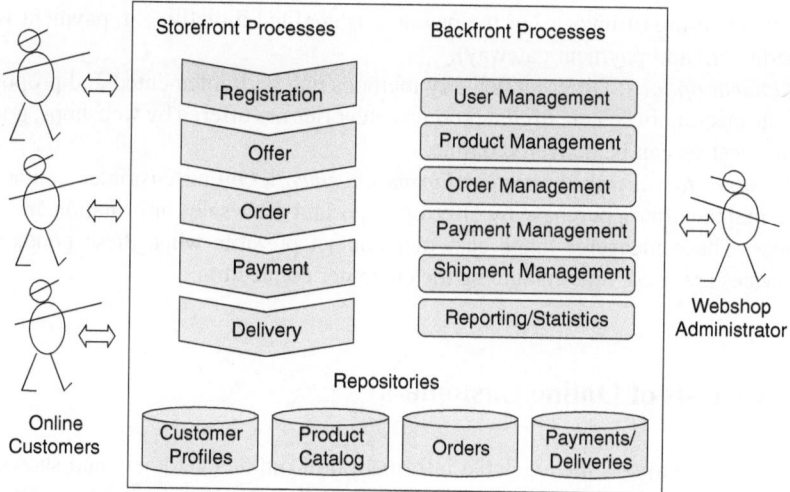

Fig. 6.2 Logical components of a webshop [83]

and services are inserted into the product catalog and the procedures for ordering, payment and purchase are specified [82].

The most important processes and repositories of a webshop, illustrated in Fig. 6.2, are [83]:

- *Registration of online customers*: A visitor of the electronic shop can find out about the products and services. Those intending to buy communicate minimum data about themselves and establish user profiles along with payment and delivery arrangements.
- *Customer profiles and customer administration*: The data on the customers is put into a database. In addition, an attempt is made to put together specific profiles based on customer behavior. This allows new, but relevant offers to be presented to the individual customer. However, the rules of communication and information desired by the user must be respected (e.g. customized push for online advertising).
- *Product catalog*: The products and services are listed in the catalog, grouped into categories so that the webshop can be clearly organized. Products may be displayed with or without prices. With individual customer pricing, a quotation is computed and specified during the drawing up of the offer, which also reflects the discount system selected.
- *Offering and ordering*: Offers can be generated and goods and services bought as needed. The electronic shopping basket or cart is used by online customers to reserve the goods and services selected for possible purchase and to preview the total price with discount.
- *Modes of payment*: If the online customer confirms his order and accepts the price and delivery arrangements, he can activate the purchase with the order button. Depending on the payment system used, this either triggers a payment process

(e.g. rendering of invoice) or the payment is credited directly (e.g. payment with credit card and payment gateway).

- *Shipment options*: Different delivery methods can be implemented and proposed to the customers. Where digital product categories are offered by webshops, goods and services can be delivered online.
- *Measures for customer relationship management*: Online customer contact is maintained after a purchase by offering important after-sales information and services. These measures make customer contact possible when these goods and services are used, thus enhancing the customer connection.

6.3 Analysis of Online Customers

This section analyzes the available information of online customers and shows, in the case of the Kiel & Co online customers, how this information can be combined in order to derive pertinent concepts and measures. For that purpose, Sect. 6.3.1 first presents how the customer data can be collected and classified based on their source and type, then Sect. 6.3.2 describes and motivates the hierarchical fuzzy classification of the Kiel & Co company.

6.3.1 Customer Profiles

The analysis of online customers compared to traditional ones has the advantage that a great deal of information about the customers' behavior is automatically logged by the online shop. The set of information describing a customer is called a customer profile (see the customer profiles repository of Fig. 6.2). Following Risch and Schubert [99], there exist different possibilities to acquire information about the customers:

- *Explicit information input*: this method, also called reactive approach, explicitly asks the customers to give information about their preferences. Preferences can be specified either by providing the users an ontology or by letting the customers rate the products or services (i.e. the preference and the ratings profiles).
- *Recording customer activity*: this second method, also called non-reactive approach, collects data from the customers' activities. Generally webshops keep a trace of the transactions (i.e. the transaction profile) and of the browsing behavior (i.e. the interaction profile) of the users. The interaction profile can then be used to derive the visiting frequency as well as the duration of the visits.
- *The usage of external data*: A third source of customer information comes from external data. For instance, back office systems (e.g. enterprise resource planning) can contain additional valuable information. If the desired information is not available inside the company, market surveys can be executed.

Table 6.1 Explicit profiles

Profile	Content
Identification profile	User name, role, contact information, personal browser settings, address, payment information, IP-address, etc.
Preference profile	Self-revealed preferences (product meta data)
Socio-economic profile	Self-categorization in predefined classes (i.e. age, gender, hobbies, etc.)
Ratings	Three types of ratings: ratings of products, of reviews and of pages
Relationships	Relationships to other users/customers
Reviews/Opinions	Plain text, images, videos and other material

Table 6.2 Implicit profiles

Profile	Content
Transaction profile	Transaction log, product purchases linked to product meta data (purchases, inquiries, payment, etc.)
Interaction profile	Click stream (pages viewed are linked to product meta data)
External data	Information procured from other sources

The information of the customer profiles, once collected, can be classified according to their source and type. Schubert [103] for instance proposes a classification scheme which distinguishes between the explicit profiles and the implicit profiles as shown in Tables 6.1 and 6.2.

6.3.2 Decomposition of the Kiel & Co Customer Profiles

Considering the online customers of the Kiel & Co company introduced at the beginning of this chapter, information have to be combined from the webshop and the ERP data in order to produce a faithful image of the customers since online and off-line channels are mostly used simultaneously. More precisely, the following information is taken into account:

- Transaction profile: Out of the transaction profile, the *turnover* and the *margin* as well as the *buying frequency* of the customers can be derived. Indirectly, the *payment delay* and the *return rate* can also be extracted.
- Interaction profile: Based on the interaction profile, the *visiting frequency* can be established.
- External data: An online survey has been executed in order to determine the *involvement frequency*.

It can be noticed that all the chosen customer information comes from the implicit profiles. The main reason underlying this fact is that implicit information contains a big amount of factual data which objectively describe the customers. For instance the transaction profile is available for all customers and the interaction profile can be used in the case of online customers. In contrast, explicit information is much scarcer and contains subjective information. For instance the socio-economic profile is a valuable information source only if all customers filled it out properly. In the context of Kiel & Co, as most transactions still occur off-line, the focus has been set on available and reliable information sources.

Based on the above-mentioned customer information, the hierarchical fuzzy classification illustrated in Fig. 6.3 can be modeled [126]. The proposed decomposition contains two main perspectives: a financial perspective expressed by the concept of profitability and a relational perspective expressed by the concept of loyalty. Since the profitability concept shows the real contribution of the customers towards the company and that the loyalty concept is a strong indicator for predicting the future behavior of the customers, the combination of these two perspectives leads to the customer lifetime value concept.

The profitability concept representing the real contribution of the customers can be built based on the gain and the service costs concepts. On the one hand, the gain concept reflects the net income generated by the customers and can be derived from the turnover and the margin information. On the other hand, the service costs concept expresses the indirect costs incurred by the business relationship and is, for the Kiel & Co company, based on the payment delay and the return rate information.

The loyalty concept, as discussed in Sect. 5.4.1, can be determined by a behavioral dimension (i.e. the repurchase rate) and an attitudinal dimension (i.e. the attachment

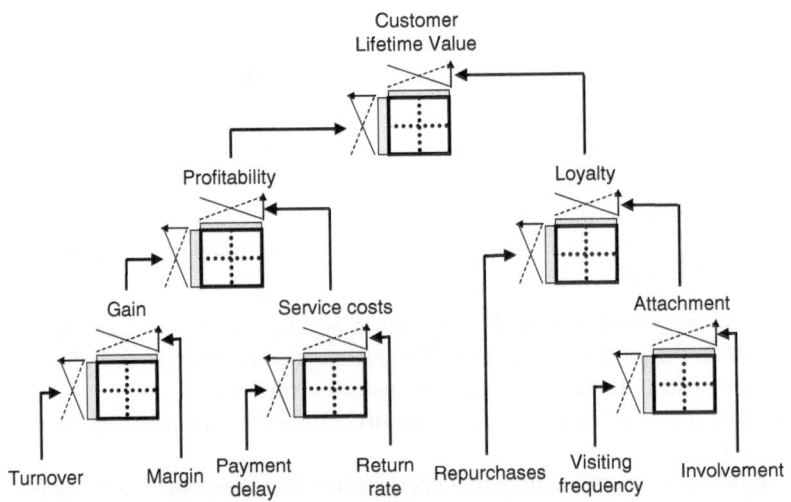

Fig. 6.3 Customer lifetime value decomposition for Kiel & Co

level). The repurchase rate can be directly obtained from the buying information. In contrast the attachment level is more complex to model. For instance, for Kiel & Co, the attachment concept has been modeled on the visiting frequency and on the involvement since users who visit an online shop frequently have a greater propensity to buy [86] and that a social involvement can lead to a virtual community that increases the trust and the attachment towards the company [98, 105].

Based on the proposed hierarchical fuzzy classification, Kiel & Co is now able to derive for every online customer his customer lifetime value, his profitability level, his loyalty grade and so on. The potential and the possible weaknesses of the classified customers can therefore be precisely determined and analyzed in all the levels of the hierarchy, based on either the resulting values of the modeled concepts or the semantics of the fuzzy classes they belong to.

6.4 Kiel & Co Case Study

This section aims at concretely illustrating the case of four typical Kiel & Co online customers in the different levels of the hierarchy of fuzzy classifications. Remember that Kiel & Co customers are either business partners or private persons and that they buy online and/or off-line. The transaction profile of the four selected customers therefore contains both online and off-line transactions. In contrast, their interaction profile (i.e. the visiting frequency) only reflects their online behavior. In order to derive the involvement attribute, Kiel & Co asked their customers to participate to an online survey for the evaluation of the firm's products and services. For all four customers, the transaction, interaction and online survey information of 2005 has been taken into account. Note that for privacy reasons the customers have been anonymized.

In order to get an overview of the context of the Kiel & Co business, some figures are presented. In 2005, Kiel & Co was selling about 1200 different products and its customer base was made up of around 6000 customers. Within the customer base, 2,400 are active customers and from these around 200 are online customers. The average turnover per customer is approximately 500€ and the best customers can have a turnover up to 50,000€. The products' margins can be very different depending on the product type but are comprised between 3 and 110 %. However margins >50 % are rare and can be found only for very specialized products.

Table 6.3 presents the selected customers as well as their respective information in order of decreasing turnover. Factory A has one of the highest turnover (58,300€) with a comfortable margin of 14 %. It visited the shop regularly (108 visits) and bought many items (65) which were paid within an average of 24 days. Constructor B has a relatively high turnover (15,073€), a short payment delay of 11 days, however its margin, visiting and buying frequencies are much lower compared to Factory A's. Furthermore Constructor B returned 6 % of its purchases. Factory C is another business customer with good visiting and buying frequencies (32 and 28) but with a relatively low turnover (6,967€), a low margin (10 %) and a long payment delay

Table 6.3 Online customers information

Customer	Turnover (€)	Margin (%)	Payment delay (days)	Return rate (%)	Buying frequency	Visiting frequency	Involvement
Factory A	58,300	14	24	0	65	108	Average
Constructor B	15,073	11	11	6	12	25	Bad
Factory C	6,967	10	26	0	28	32	Bad
Customer D	673	23	0	0	11	12	Good

(26 days). Finally, Customer D, which is the only private customer in our test set, bought items for a value of only 673€. His margin is however very high compared to the business customers (23 %). Since private customers have to pay in advance, Customer D always has a payment delay of 0 day.

In contrast to all the other attributes, the involvement information is a qualitative attribute. As mentioned above, this attribute is based on the response to an online survey and is therefore derived from a general judgement as follows: customers who did not respond are given a 'Bad' value; customers having partially filled the form are rated as 'Average' whereas those who fully completed the survey have a 'Good' involvement.

6.4.1 Decomposition of the Profitability Concept

The profitability of a customer is the difference between the revenue he generates and the costs he causes. The sub-classification profitability is then defined by the concepts gain and service costs. The gain concept represents the profits of a customer's purchases. This concept can be easily calculated based on the attributes turnover and margin derived from the transaction profile (see Fig. 6.4).

The definition of this fuzzy classification is achieved in three steps:

- First, the pertinent domain ranges have to be determined for the turnover and the margin attributes. Since the best Kiel & Co customers have a turnover of about 50,000€, the domain of the attribute turnover is limited to the interval 0 to 50,000. This means that any customer having a greater turnover is classified with the maximum allowed value (i.e. 50,000 in this case). In a similar way, the margin domain range has been defined between 0 and 50 % since most products lay in this interval.
- In a second step, based on the chosen domain definitions, the membership functions for the linguistic terms have to be defined. In this case study, all the membership functions are linear in order to be able to distinguish between all customers within the attributes' domain.

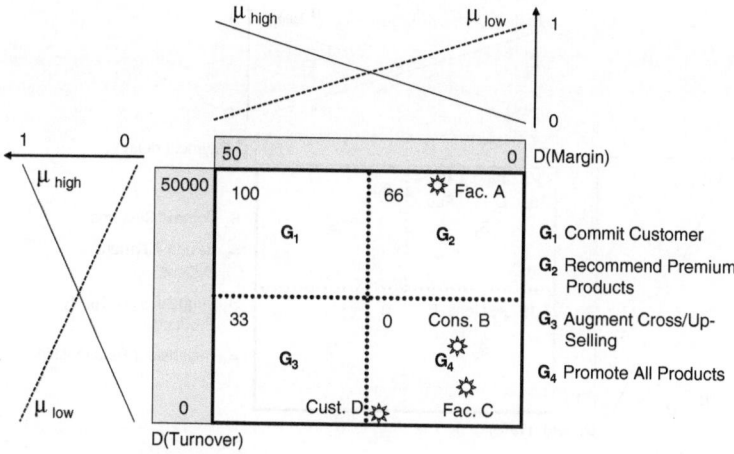

Fig. 6.4 Gain sub-classification

Table 6.4 Calculation of the gain concept

Customer	Turnover	Margin	Gain
Factory A	58,300	14	79
Constructor B	15,073	11	35
Factory C	6,967	10	26
Customer D	673	23	21

- Finally, the third step consists of assigning grades, i.e. gain grades, to the classes in order to be able to compute a gain value for each customer. For all the fuzzy classifications, the grades ranges from 0 to 100 and have been assigned linearly from the worst to the best class.

Table 6.4 recapitulates the attributes values for turnover and margin and also indicates the gain value calculated by means of the personalized value introduced in Definition 5.1. Due to its maximal turnover and its good margin Factory A easily leads the gain concept with a personalized value of 79. The two other business partners are logically far behind, Constructor B being nevertheless better placed due to a much better turnover and a slightly bigger margin. Despite his tiny turnover, Customer D falls just behind Factory C since his excellent margin largely compensates his turnover.

The service costs concept reflects the indirect costs that the customers can generate by returning products and by not paying on time (see Fig. 6.5). The attribute return rate is the percentage of returned products; its domain has been limited to [0, 10] as 10 % is already a non-acceptable return rate for the Kiel & Co company. The payment delay attribute represents the average delay in days of a customer after the product has been shipped and has been defined over a domain ranging from 0 to 60.

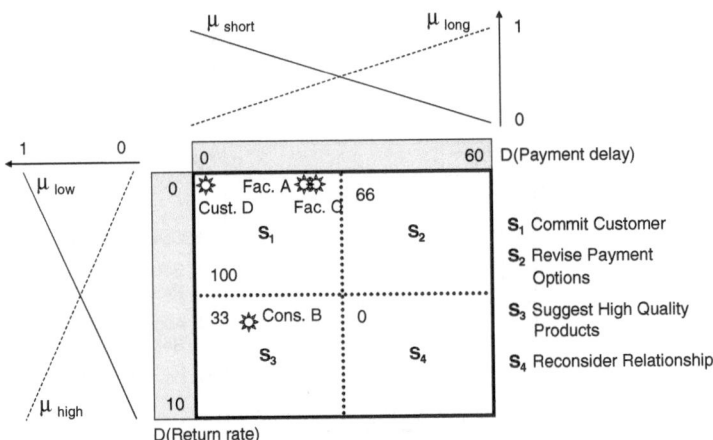

Fig. 6.5 Service costs sub-classification

Table 6.5 Calculation of the service costs concept

Customer	Return rate	Payment delay	Service costs
Factory A	0	24	85
Constructor B	6	11	53
Factory C	0	26	84
Customer D	0	0	100

Note that the payment delay is typically an attribute which can be assessed using an overall judgment (i.e. a qualitative attribute) instead of numerical values.

Table 6.5 shows the service costs grades obtained based on the return rate and the payment delay values. Note that in this case study 100 always refers to the best grade. In this classification, Customer D gets the best grade since he never returned articles and had to pay in advance according to the Kiel & Co policy. With also no return and an average payment delay, Factory A and C obtain the good personalized values of 85 and 84 respectively. Despite its relatively short payment delay, Constructor B lags behind due to its high return rate.

The gain and service costs grades calculated in their respective fuzzy classification can now be used to define the profitability sub-classification (see Fig. 6.6). Since the grades for the gain and service costs concepts have both been defined on the interval [0, 100], their domain definitions are logically identical.

Table 6.6 recalls the gain and service costs values and gives the profitability grades. Having high gain and service costs grades, Factory A is largely ahead with a grade of 71. More surprising is the second place of Customer D with a personalized value of 56. Due to his outstanding service costs performance, he can compensate his modest gain. Constructor B and Factory C get similar grades, Constructor B being ranked last due to its service costs problems.

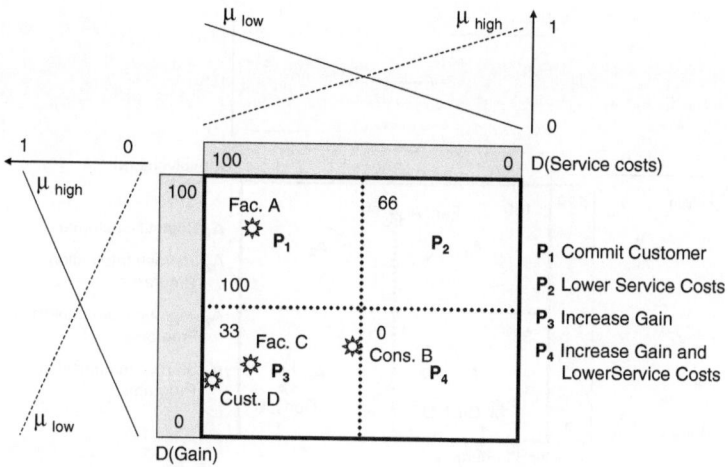

Fig. 6.6 Profitability sub-classification

Table 6.6 Calculation of the profitability concept

Customer	Gain	Service costs	Profitability
Factory A	79	85	71
Constructor B	35	53	44
Factory C	26	84	48
Customer D	21	100	56

6.4.2 Decomposition of the Loyalty Concept

The attachment concept can be defined in terms of visiting frequency and involve-ment as shown in Fig. 6.7. The involvement has a deeper meaning than the visiting frequency as sharing experiences denotes a greater attachment than just browsing an online shop. The involvement of the customers towards a company can be measured based on product ratings, forum entries, online survey and all kinds of interactions showing an implication of the customers. Since Kiel & Co did not have enough product ratings available in 2005, the response rate to an online survey has been considered.

For the visiting frequency, the domain has been limited to the interval [0, 120]. This definition comes from the observation that the best customers generally visit the shop every second working day. The results of the attachment concept are given in Table 6.7. Thanks to its very high visiting frequency and having partially filled out the survey, Factory A gets the good grade of 70. Once again, Customer D arrives second with a score of 50 because he fully replied the questionnaire. Also, Constructor B and Factory C lay behind, Factory C having a short lead due to its higher visiting frequency.

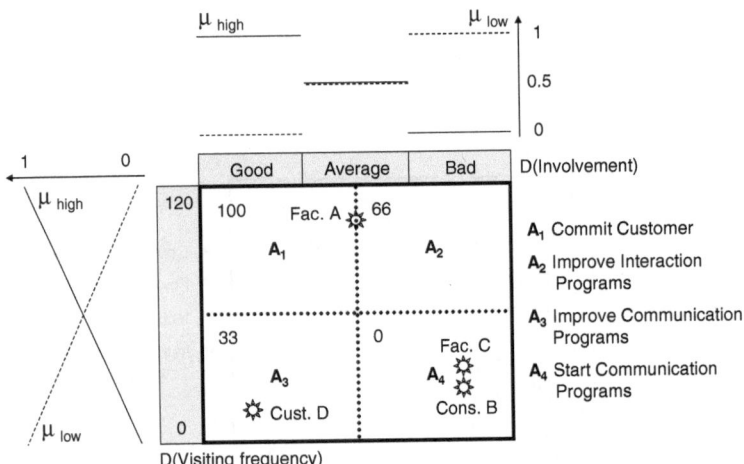

Fig. 6.7 Attachment sub-classification

Table 6.7 Calculation of the attachment concept

Customer	Visiting frequency	Involvement	Attachment
Factory A	108	Average	70
Constructor B	25	Bad	22
Factory C	32	Bad	25
Customer D	12	Good	50

The loyalty sub-classification, shown in Fig. 6.8, is defined by the dimensions buying behavior and attachment as proposed by Harrison [48]. In contrast to the attachment, the buying frequency or repurchases attribute can be directly derived from the order information. For Kiel & Co, the domain of the attribute buying frequency is the interval [0, 60] since good customers place around 60 orders per year.

The results of the loyalty sub-classification are shown in Table 6.8. Having an excellent buying frequency and a good attachment Factory A remains in the first position with a loyalty grade of 86. Then Factory C recovers the second place with a score of 43 having a higher buying frequency than Customer D and Constructor B. Customer D still has a slight advantage over Constructor B thanks to his attachment grade.

6.4.3 Customer Lifetime Value Classification

The top fuzzy classification shown in Fig. 6.9 expresses the concept of customer lifetime value (see Sect. 5.4.2). It is based on the already defined concepts of customer profitability and loyalty. The profitability concept, which covers the finan-

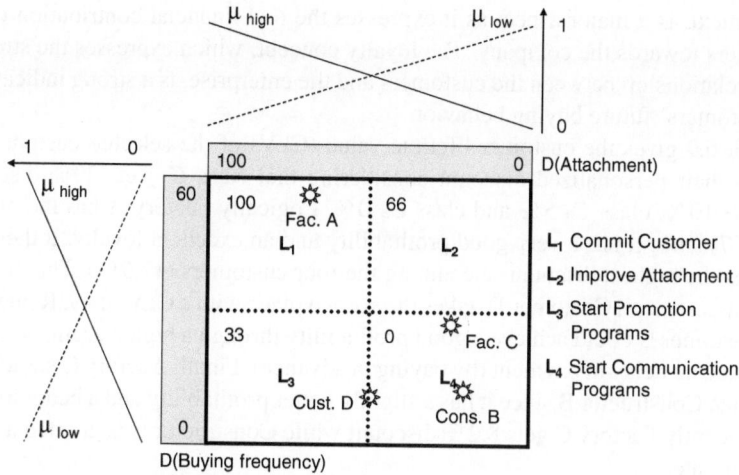

Fig. 6.8 Loyalty sub-classification

Table 6.8 Calculation of the loyalty concept

Customer	Buying frequency	Attachment	Loyalty
Factory A	65	70	86
Constructor B	12	22	30
Factory C	28	25	43
Customer D	11	50	35

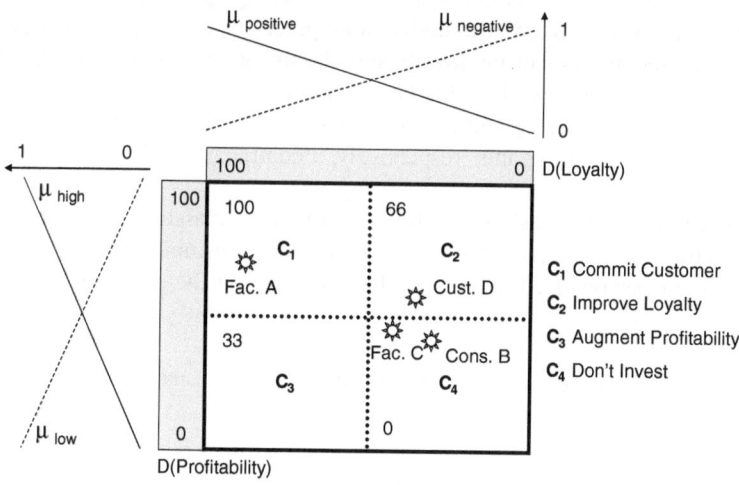

Fig. 6.9 Top fuzzy classification expressing the customer lifetime value

cial context, is a major aspect as it expresses the real financial contribution of the customers towards the company. The loyalty concept, which expresses the strength of the relationship between the customers and the enterprise, is a strong indicator of the customers' future buying behavior.

Table 6.9 gives the customer lifetime value (CLV) of the selected customers as well as their personalized discount considering that class C_1 gets 20 % discount, class C_2 10 %, class C_3 5 % and class C_4 0 %. Logically Factory A has the highest CLV (67) since it has a very good profitability and an excellent loyalty. It therefore receives the highest discount rate among the four customers (12.5 %). The outsider of this assessment, Customer D, takes the second place with a CLV of 49. Remember that Customer D could achieve a good profitability through a high margin, no article return and no delay in payment (by paying in advance). Finally Factory C has a better CLV than Constructor B since it has a slightly better profitability and a better loyalty. Consequently Factory C gets 8.2 % discount while Constructor B is granted a 7.2 % discount rate.

The moral of this case study may be not to draw direct conclusions whenever some customer figures are available. As previously discussed for Customer D, a small turnover does not always imply a non profitable customer since all the other aspects of this customer need to be considered. The fuzzy classification approach therefore allows a fair and transparent judgment of the customers in all the levels of the hierarchy. What has not yet been emphasized is that for each customer in each fuzzy classification, the semantics of the classes he belongs to can be used in order to detect his potential weaknesses and also the means to improve his value. Consider for instance that the management of Kiel & Co wants to develop Customer D as he has a good potential. By looking at the top fuzzy classification space in Fig. 6.9, it becomes clear that Customer D has the highest belonging to class C_2 with semantics 'improve loyalty'. In order to derive more precisely where the loyalty problem comes from, the analysis of the loyalty sub-classification is necessary. Clearly the main problem of Customer D is his buying frequency (see Fig. 6.8) and, as he is lying between the classes L_3 and L_4 with semantics 'start promotion programs' and 'start communication programs' respectively, a combination of both strategies can be undertaken.

Furthermore, all the possibilities offered by the fuzzy classification approach, like the marketing campaign planning and optimizing, the customer monitoring and the calculation of personalized values, can be performed at the different levels of the

Table 6.9 Customer lifetime value and personalized discount calculation

Customer	Profitability	Loyalty	CLV	Discount (%)
Factory A	71	87	67	12.5
Constructor B	44	30	42	7.2
Factory C	47	43	47	8.2
Customer D	56	35	49	8.5

hierarchy. For instance if a loyalty campaign has to be launched, the targeting and the controlling of the campaign should take place in the loyalty sub-classification and not at the top level. It is also possible to derive several concepts from the same fuzzy classification with different metrics (i.e. the grades assigned to the classes) to precisely model specific aspects. In the top fuzzy classification, for example, the customer lifetime value and the personalized discount have been calculated using two different metrics.

To conclude this section, it is important to stress once more that the modeling of a hierarchy of fuzzy classifications, whatever application area of consideration, is context dependent and subjective. Context dependent means that the application domain, the type of the customer base and the size of the enterprise strongly influence the results. Subjective means that given the same context, two managers would probably achieve different models. In the field of data analysis, another restriction is the quantity and the quality of available data. Therefore the modeling of a hierarchical fuzzy classification, i.e. the selection of the qualifying attributes, the introduction of the equivalence classes, the definition of the membership functions, the decomposition process and the specification of the personalized values are important design issues which require the participation of the management, marketing specialists and database architects.

Chapter 7
fCQL Toolkit

Since the concept of the fuzzy classification has been explained in Chap. 3, the pertinence of the fuzzy classes has been illustrated in Chap. 5 and one possible application domain has been discussed in Chap. 6, this chapter finally demonstrates the architecture, the functionalities, the user interface as well as the relational database schema of the fCQL toolkit.

For this purpose, Sect. 7.1 presents the general architecture of the fCQL toolkit, Sect. 7.2 shows the toolkit's functionalities and interface whereas Sect. 7.3 depicts the database schema of the additional meta-tables necessary to store the fuzzy classifications definitions.

7.1 Architecture of the fCQL Toolkit

As explained in Chap. 3, the fuzzy classification approach is achieved by extending the relational database schema. This extension consists of meta-tables added to the system catalog. The meta-tables contain all the definition information of the fuzzy classifications, i.e. the definition of the linguistic variables and terms, the description of the classes and all the meta-information regarding the membership functions and the decomposition process (see Sect. 7.3). As shown in Fig. 7.1, the meta-tables are clearly independent from the business data so that the insertion, the modification and the removal of the fuzzy classification using the fCQL toolkit do not affect the business data as well as the consistency of the database. Furthermore, the use of the fCQL toolkit does not require any transformation nor any migration of the existing data, only the creation of a view is needed in the case the qualifying attributes have to be aggregated from different database tables.

The architecture of the fCQL toolkit shown in Fig. 7.1 illustrates the interactions between users, the fCQL toolkit and the relational database management system (RDBMS). The fCQL toolkit consists of an additional layer above the relational database system; this characteristic makes fCQL independent of underlying database systems and thus enables fCQL to operate with every database product. It also

© Springer International Publishing Switzerland 2015
N. Werro, *Fuzzy Classification of Online Customers*, Fuzzy Management Methods,
DOI 10.1007/978-3-319-15970-6_7

Fig. 7.1 Architecture of the fCQL toolkit

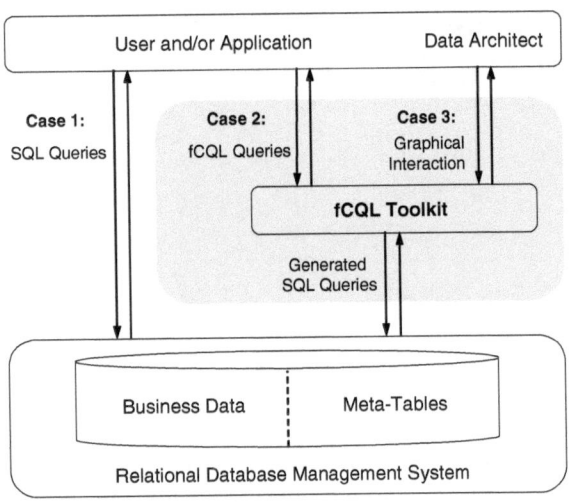

implies that the user can always query the database with standard SQL commands (see Case 1). This is a major advantage since companies are often refrained from implementing new technologies due to compatibility issues.

Additionally, the fCQL toolkit allows the user to formulate unsharp classification queries (see Case 2). These queries are then analyzed and translated into corresponding SQL statements for the RDBMS. In order to generate the fuzzy classification results, the toolkit accesses the business data as well as the meta-data and computes the membership degrees of the classified elements in the different classes. The classification results are finally displayed to the user or returned to the application. Before querying the fCQL toolkit, the data architect has to define the fuzzy classification (see Case 3). This task can be done graphically with the help of a wizard which goes step by step over the definition process (see Sect. 7.2).

The fCQL toolkit is a standalone application implemented in Java which is actually the leading object-oriented programming language [36]. The choice of Java among other popular technologies has been motivated by the abundance of the Java platform, the support of the most recent technologies and the large Java community which provides open source developing and running environments. Another important advantage of using the Java language is the portability of the code, meaning that Java programs can be run on all main operating systems, e.g. Windows, Mac OS, Unix, Linux, Solaris, etc. Java applications can also easily establish connections to RDBMS using a JDBC driver often delivered by the DBMS vendor itself. The fCQL toolkit contains drivers and adapters for the two main open source database systems namely MySQL[1] and PostgreSQL.[2] Finally, Java offers many graphic libraries which enable an ergonomic and user-friendly user interface.

---

[1] Available at http://www.mysql.com (last visited 07/Oct/2008).

[2] Available at http://www.postgresql.org (last visited 07/Oct/2008).

7.2   User Interface—A Guided Tour

This section serves as a short guided tour to provide the reader a first impression about the fCQL toolkit's capabilities. The main stages which deserve a look are the database connection, the data analysis panels, the fuzzy classification definition wizard, the fuzzy querying process and the results evaluation based on the classification example provided by the fCQL toolkit.

The main window of the fCQL toolkit shown in Fig. 7.2 consists of three main panels, i.e. the one-dimensional data analysis, the two-dimensional data analysis and the fuzzy classification panel. The two first panels allow the user to perform some basic data analysis in order to help him defining the fuzzy classifications. Once a fuzzy classification has been implemented, the user can see its definition in the info sub-panel (see Fig. 7.2) and execute fuzzy classification queries in the query sub-panel.

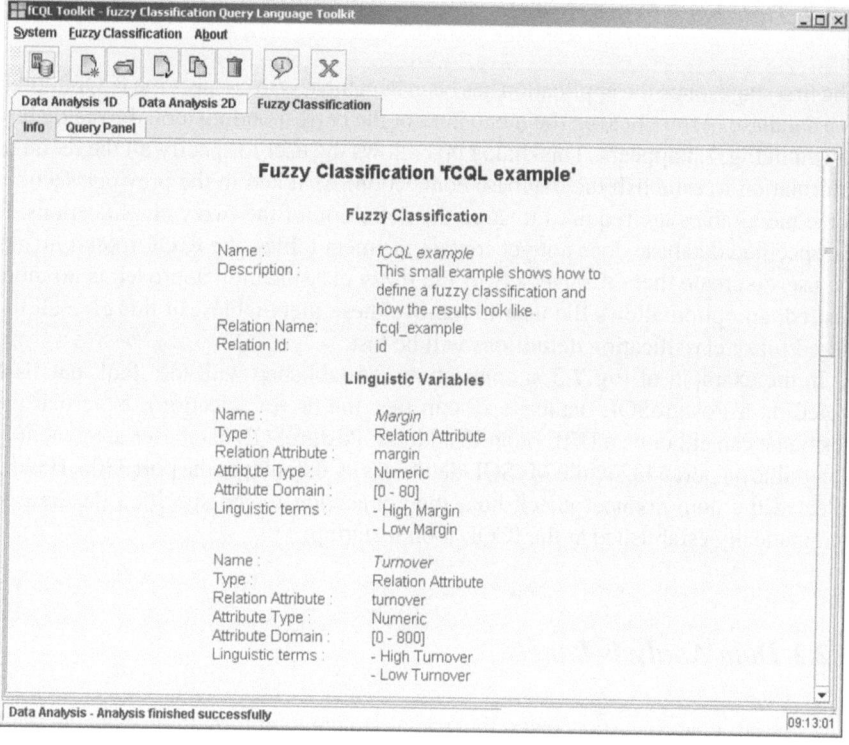

Fig. 7.2 Main window of the fCQL toolkit

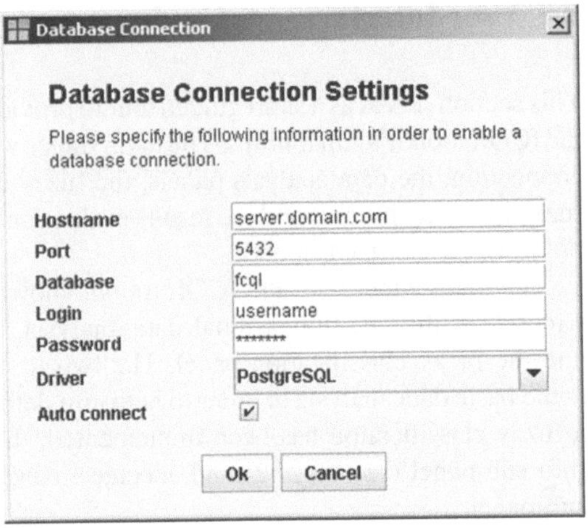

Fig. 7.3 Dialog box for
connecting a database

7.2.1 Database Connection

The first stage, once the application has been launched, is to connect the fCQL toolkit
to a database. After clicking the menu item or the corresponding icon the dialog box
shown in Fig. 7.3 appears. This dialog box allows the user to specify all the required
information to establish the database connection. As noted in the previous section,
some meta-tables are required to store the definition of the fuzzy classifications. If
the specified database does not yet contain the meta-tables, the fCQL toolkit invites
the user to create them. Conversely, if the fuzzy classification approach is no more
desired, an option allows the user to remove these meta-tables. In this case all the
stored fuzzy classification definitions will be lost.

In the example of Fig. 7.3, a connection is established with the 'fcql' database
stored on a PostgreSQL database system (see the driver selection). Note that the
hostname can either be a URL or an IP address. PostgreSQL databases are generally
accessible on port 5432 while MySQL databases by default use the port 3306. Having
selected the auto connect check box, the connection to the specified database is
automatically established at the fCQL toolkit startup.

7.2.2 Data Analysis Panels

The second stage consists of extracting some elementary information about the available attributes of the elements under consideration. This information is most useful
to determine the fuzzy classification definition, i.e. the selection of the qualifying
attributes, the specification of the attributes' domain, the choice of the equivalence
classes as well as the shape of the membership functions. For this purpose, the fCQL
toolkit offers two basic data analysis features, a one-dimensional data analysis which

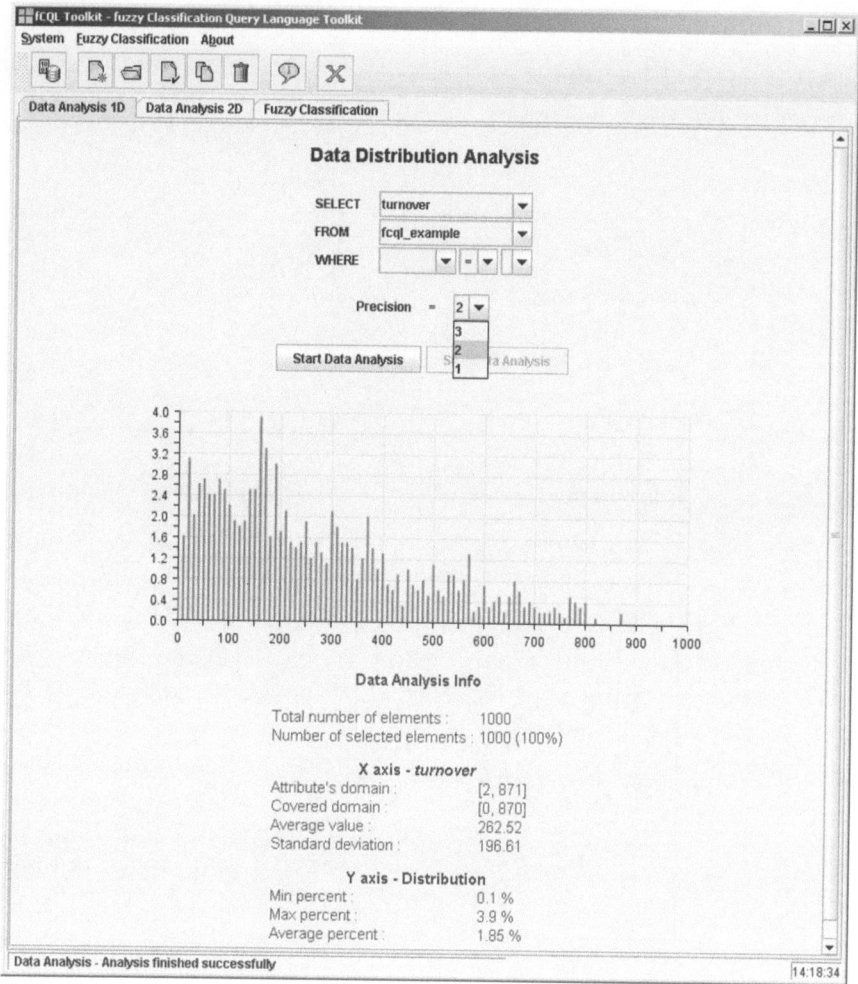

Fig. 7.4 One-dimensional data analysis

shows the distribution of the data on the attribute's domain and a two-dimensional data analysis which visualizes the elements in a two-dimensional space.

The one-dimensional data analysis panel illustrated in Fig. 7.4 analyzes the specified attribute based on a distribution graph and some statistical information. The X-axis of the distribution graph represents the values of the attribute and the Y-axis gives the percentage of elements having the same value. Considering the fCQL example relation containing 1,000 customers with attributes turnover and margin, the one-dimensional data analysis panel gives the following statistical information:

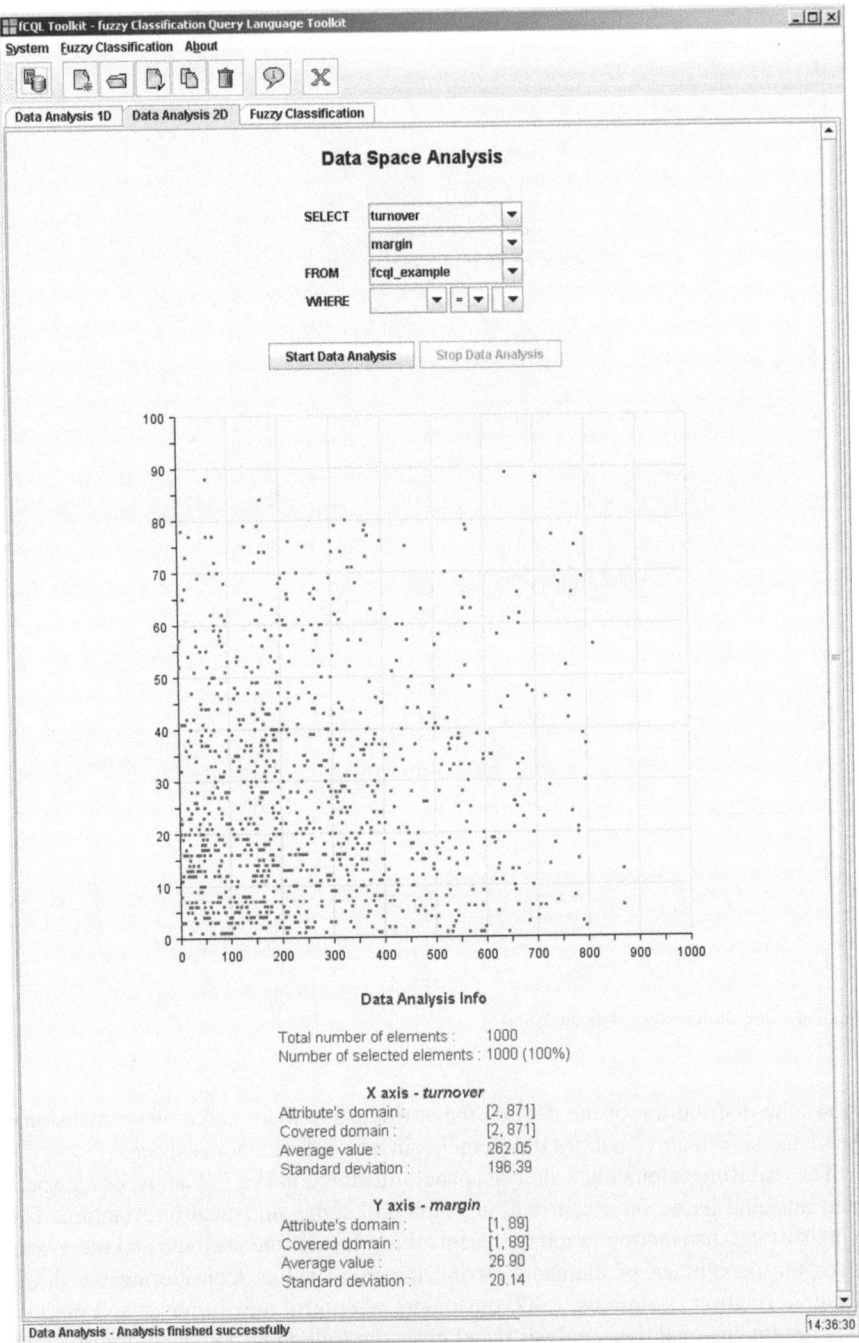

Fig. 7.5 Two-dimensional data analysis

- *Data analysis information*: The total number of elements in the data set and the number of elements complying with the selection condition (if any). In this case all customers are considered.
- *X axis*: The attribute's domain considering all the elements, the covered range, the average value and the standard deviation of the selected elements. In this example, the turnover attribute's domain ranges from 2 to 871. But since the precision has been set to value 2 (see below), the turnover values are rounded and the covered domain ranges therefore from 0 to 870. The average turnover and the standard deviation are 262.52 and 196.61 respectively.
- *Y axis*: The minimum, the maximum and the average percentages of elements having the same values. Due to the aggregation of turnover values to tens, the minimum percentage is 0.1, the maximum percentage is 3.9 and the average percentage is 1.85.

In some cases where the data precision is very high, the distribution information does not deliver valuable information. For this reason, a precision parameter is available below the SQL command allowing the user to aggregate the attribute's values. For example, the customers' turnover does not need the unitary precision. By adjusting the precision parameter, all panel's information are updated with the chosen precision and tendencies can be better detected. Note that categorical attributes can be analyzed the same way except that the average value, the standard deviation and the precision parameter are not available.

In the two-dimensional data analysis panel, two attributes can be specified forming a two-dimensional space where the selected elements can be visualized (see Fig. 7.5). By combining qualifying attributes it provides a good preview of the distribution of the elements in the potential classification spaces. The two-dimensional data analysis panel provides statistical information about the number of elements and selected elements as well as the attribute's domain, the covered domain, the average value and the standard deviation for the two given attributes. Note that the two-dimensional data analysis panel can also be used to detect linear relationships between attributes.

7.2.3 Fuzzy Classification Definition Wizard

The third stage relates to the definition of a new fuzzy classification or to the editing of an existing fuzzy classification. For that purpose, the fCQL toolkit proposes a graphical wizard which guides the user through the definition process. This process can be decomposed into six steps. The first step, depicted in Fig. 7.6, relates to the basic information concerning the fuzzy classification, i.e. the name, the description, the relation or the view holding the qualifying attributes and a relation identifier. The relation identifier is necessary for combining relational attributes and composed attributes in order to build fuzzy classification hierarchies.

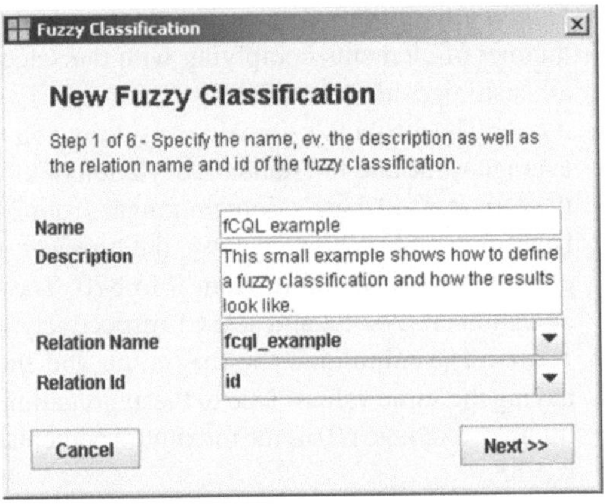

Fig. 7.6 Basic fuzzy
classification information

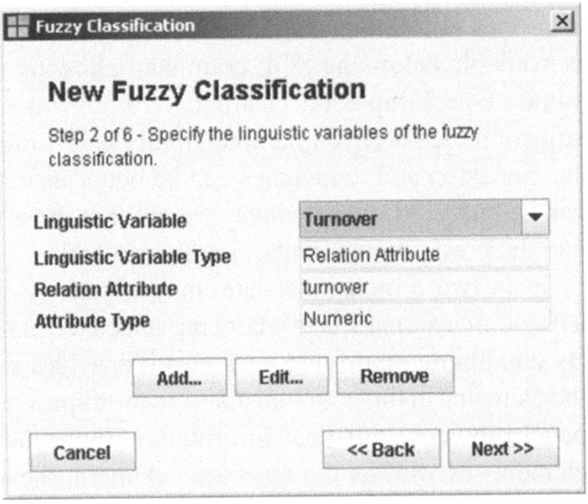

Fig. 7.7 Linguistic variables
selection and specification

In the second step the user is able to specify the pertinent attributes, i.e. the linguistic variables, for the fuzzy classification. The user can add, edit or remove linguistic variables as shown in Fig. 7.7.

When adding a linguistic variable, its name and its type (i.e. relational or composed attribute) have to be indicated. Note that the name of the linguistic variable is independent of the attribute's name. At this point, different dialog boxes appear depending on whether the linguistic variable is based on a relational attribute or on a composed one (see Fig. 7.8). In the case of relational attributes, the name and the type (i.e. numeric or categoric) of the attribute have to be specified. For composed attributes the underlying fuzzy classification as well as the concept generating the composed attribute have to be defined. Note that in a given fuzzy classification each attribute, either relational or composed, can only be selected once.

For relational attributes of type numeric as for composed attributes, both dealing with numerical values, the considered domain has to be determined (see Fig. 7.9). The considered domain can be broader or smaller than the actual attribute's domain. In the

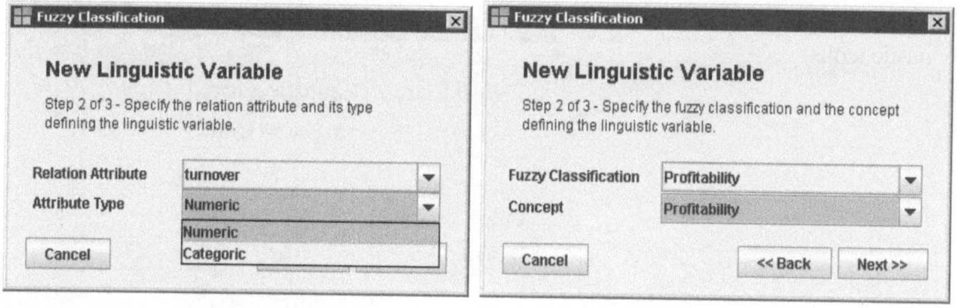

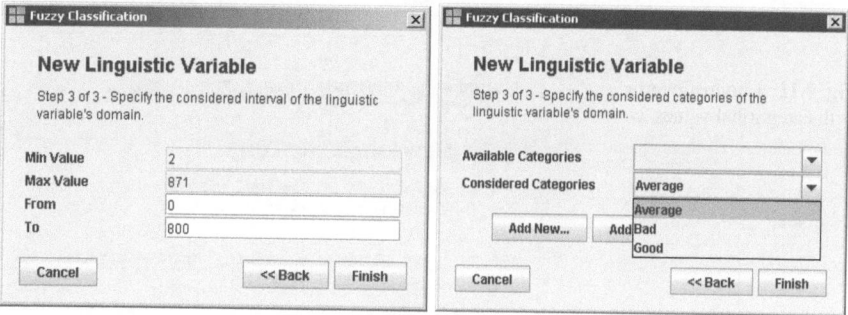

Fig. 7.8 Linguistic variable's types

Fig. 7.9 Linguistic variable's domain and categories

latter case, values exceeding the considered domain are automatically brought back to the domain's boundaries. For relational attribute of type categorical, the possible categories have to be listed (see Fig. 7.9). Note that the definition of the linguistic terms relies on the categories of the linguistic variable. Therefore elements holding missing categories will not be considered during the fuzzy classification querying.

Once the linguistic variables have been defined, fuzzy equivalence classes, i.e. the linguistic terms, can be characterized in the third step. For each linguistic variable, the user can add, edit or remove linguistic terms as illustrated in Fig. 7.10.

Here once again, a distinction has to be done between linguistic terms of a linguistic variable based on numerical or categorical values. Linguistic terms dealing with numerical values are most suited since they can be defined using a membership function spanning the considered domain of the linguistic variable. In contrast, linguistic terms relying on categorical values require a membership degree assignment for each possible category defined in the linguistic variable (see Fig. 7.11).

An important feature of the fCQL toolkit wizard is its membership function editor which allows the user to graphically define and edit the membership function of the numerical linguistic terms as shown in Fig. 7.12. As a result the user does not have to deal with complicated mathematical functions and can easily define the appropriate curve of the membership function [89].

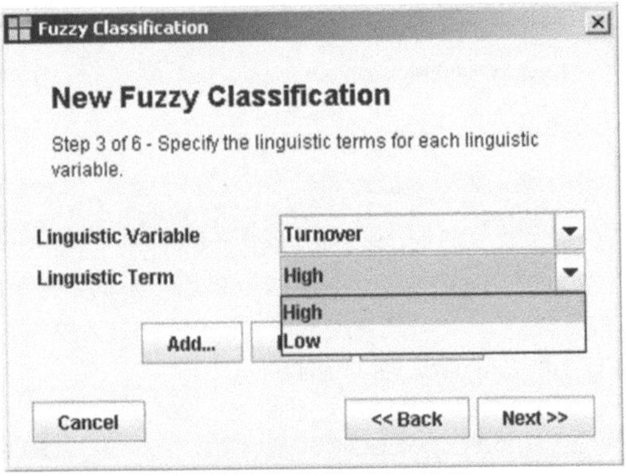

Fig. 7.10 Defining
linguistic terms

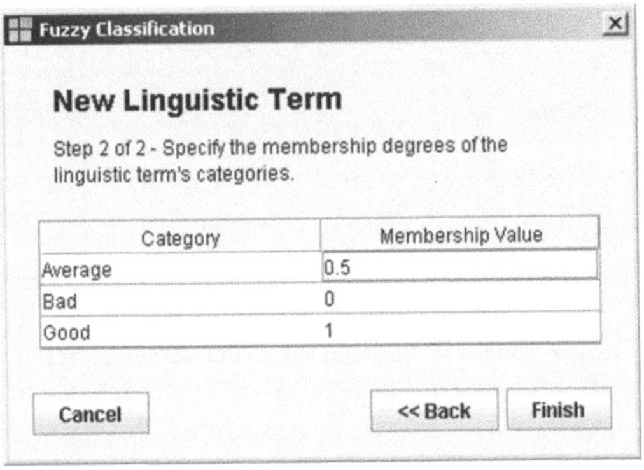

Fig. 7.11 Linguistic term
with categorical values

In order to define continuous membership functions, the fCQL toolkit offers two different function shapes. The first shape is a linear function from which the well known triangular and trapezoidal functions can be derived. The second shape, which is more sophisticated, is the S-shaped function. This parameterized curved function was developed by Dombi [28] to enable the representation of human concepts. By combining the two available membership function shapes, a wide range of membership functions can be generated. Note that it is furthermore possible to add new types of functions for particular needs.

In the fourth step, the final fuzzy classes have to be described. The definition of m linguistic variables with $n_{1..m}$ linguistic terms leads to a classification space with $\prod_{i=1}^{m} n_i$ fuzzy classes. The fuzzy classes are automatically generated and are given a default name based on the generation sequence. Based on the linguistic terms defining the fuzzy classes, the user only has to edit the fuzzy classes by giving them an appropriate name and description (see Fig. 7.13). Note that adding or removing linguistic variables or terms implies a new classification space, i.e. new fuzzy classes are generated.

As explained in Chap. 5, by assigning grades to the fuzzy classes it is possible to derive new concepts. The concepts and the associated grades can be defined in

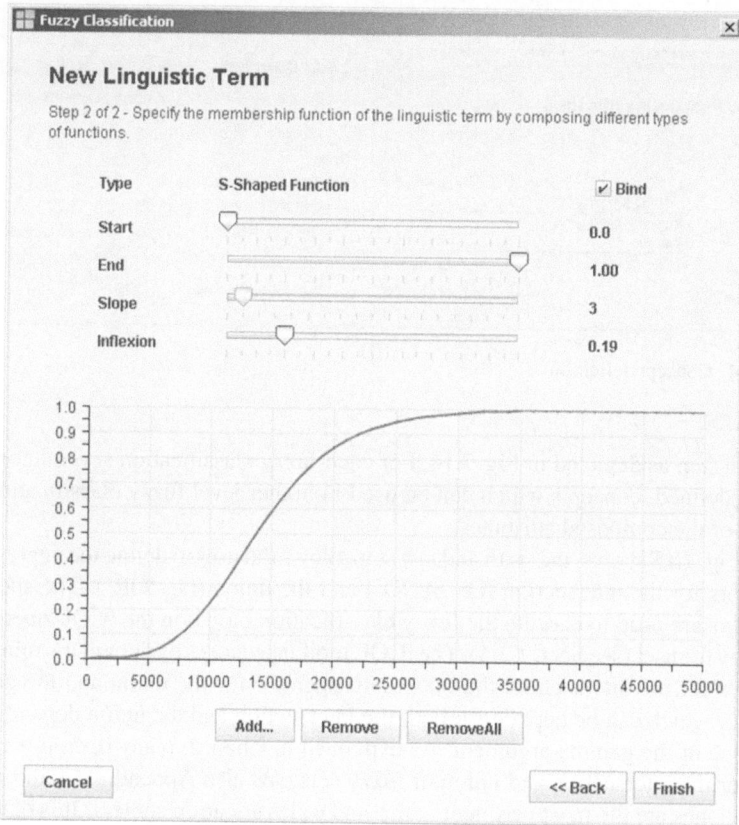

Fig. 7.12 Membership function editor

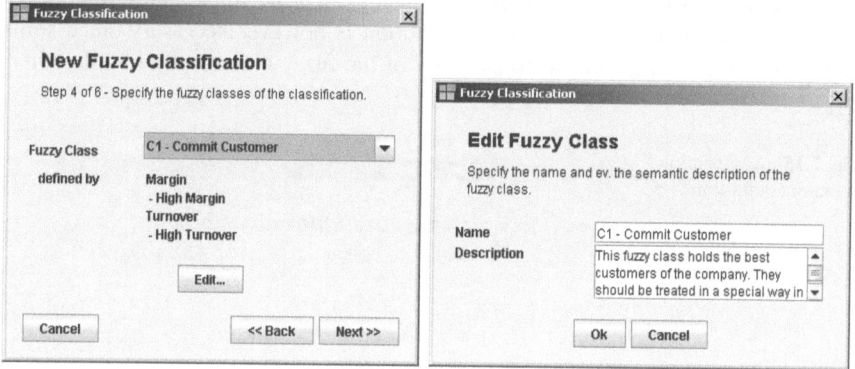

Fig. 7.13 Description of the fuzzy classes

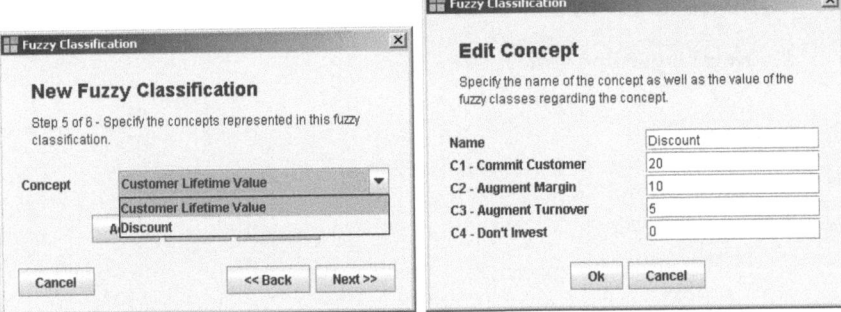

Fig. 7.14 Concept definition

the fifth step as depicted in Fig. 7.14. For each fuzzy classification several concepts can be defined, concepts which can be used in higher level fuzzy classifications by means of the composed attributes.

As Fig. 7.15 shows, the sixth and last step allows the user to define the aggregation operators for the intersection (i.e. 'AND') and the union (i.e. 'OR'). The specified operators are used to execute the fuzzy classification based on the fCQL query provided by the user (see Sect. 7.2.4). The fCQL toolkit proposes by default the minimum operator for the intersection, the maximum operator for the union and the gamma operator which can be applied for both the intersection and the union depending on the value of the gamma argument. As explained in Chap. 2, many operators can be used for the intersection and union of fuzzy sets (see also Appendix A); the implemented ones are the most pertinent in the fuzzy classification context. It is of course possible to add further operators in order to suit specific needs.

Finally, by clicking the finish button, the fuzzy classification definition is stored in the meta-tables (see Sect. 7.3). The user can at anytime edit existing fuzzy classifications and modify their definition. Caution is however necessary since some modifications do not affect the consistency of the fuzzy classification while others

Fig. 7.15 Aggregation operators definition

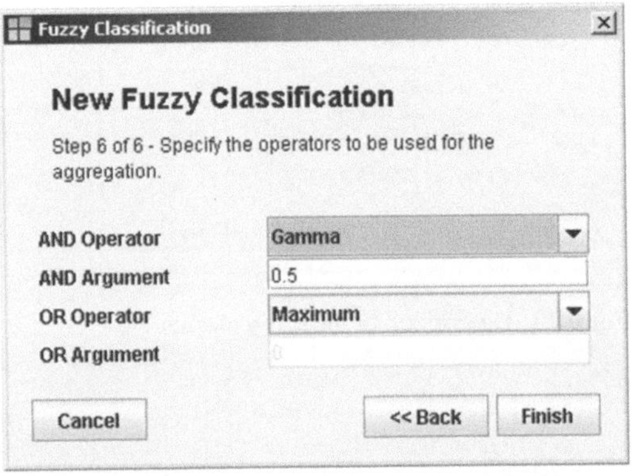

do. For instance, changing the name and the description of the different elements, modifying the shape of the membership functions and adapting the classes' grades do not modify the main structure of the fuzzy classification. In contrast, operations like changing the attributes' type, redefining the considered domains of the qualifying attributes and adding or removing linguistic variables and terms imply further adjustments, i.e. the redefinition of the membership functions and/or of the fuzzy classes.

7.2.4 Fuzzy Querying Process

The fourth and main stage of the fuzzy data analysis is the fuzzy querying process. Querying a fuzzy classification is very intuitive as all the fuzzy classification components have already been defined with the help of the definition wizard. Furthermore the fuzzy classification queries can be graphically defined by clicking the dynamic querying panel. The fCQL syntax, which has been introduced in Chap. 3, and the querying results are concretely demonstrated in this subsection.

Generally the fuzzy classification is executed for all the elements in all the fuzzy classes in order to retrieve the membership values of the elements in the final fuzzy classes and the related concept values. In this case, it is sufficient to start the query analysis without specifying any *where*, *with* nor *alpha* condition (see Fig. 7.16). Note that the *classify* clause does not impact the classification results since it only specifies which attributes are displayed in the results table. By default, the attributes of the linguistic variables are selected. Logically the *from* clause cannot be edited since it is specified in the fuzzy classification definition.

Just as in standard SQL queries, the user might want to restrict the data set under consideration by specifying a selection condition in the *where* clause. For instance,

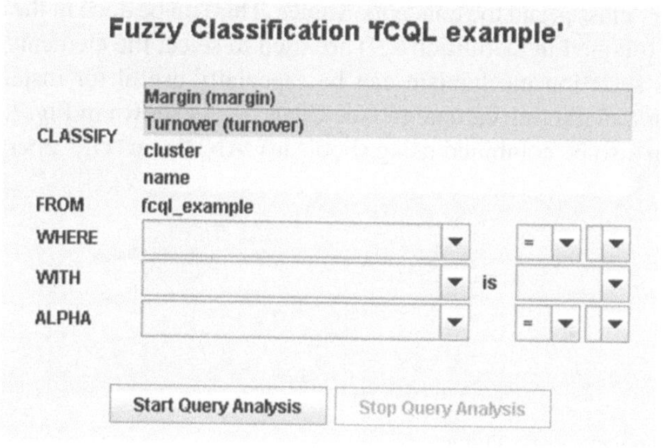

Fig. 7.16 Fuzzy querying panel

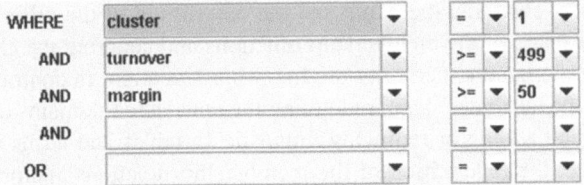

Fig. 7.17 Composed *where* condition

customers could be selected based on clustering information as well as on a turnover and margin threshold as shown in Fig. 7.17. As soon as a valid condition is entered, the fCQL toolkit automatically and recursively adds blank predicates allowing to express composed *where* conditions with the (boolean) 'AND' and 'OR' operators. Note that the *where* clause is identical in the fuzzy querying panel as in the data analysis panels.

Based on the fuzzy classification definition, the user can specify a classification condition in the *with* clause. This condition can be specified either on a class level or on a linguistic term level. By giving a classification condition, the user considers only one fuzzy class which can be a predefined final fuzzy class or a new fuzzy class. Generating a new fuzzy class has to be done with caution since its semantics has to be well understood in order to not misinterpret the classification results. Here again the querying panel adds new blank predicates in order to enable the fCQL syntax (see Fig. 7.18). In contrast to the *where* clause, the 'AND' and 'OR' operators of the *with* condition are the aggregation operators defined during the fuzzy classification wizard. If no *with* condition has been specified, the 'AND' aggregation operator is also used by the fCQL toolkit in order to calculate the belonging of the elements in the predefined final classes.

Finally, the user has the possibility of selecting elements, not based on the attributes' values, but on the fuzzy classification results, i.e. the belonging in the different fuzzy classes and the concepts' values. This can be done in the *alpha* clause since α-cuts (defined in Definition 2.9) are used to select the elements in the fuzzy classes. This selection mechanism can be especially useful for instance to target customers for a marketing campaign (see Chap. 5). As shown in Fig. 7.19 the alpha condition can also be combined using (boolean) 'AND' and 'OR' operators.

WITH	class		is	C1 - Commit Customer	
OR	Turnover		is	High	or	Low
AND	Margin		is	Low	or
OR			is		

Fig. 7.18 Composed *with* condition

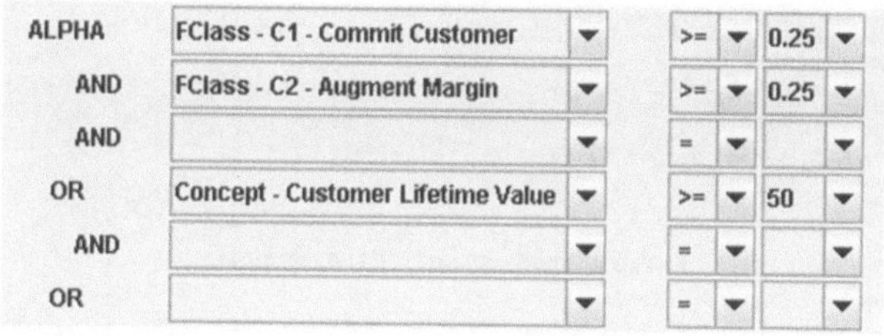

ALPHA	FClass - C1 - Commit Customer	▼	>=	▼	0.25	▼
AND	FClass - C2 - Augment Margin	▼	>=	▼	0.25	▼
AND		▼	=	▼		▼
OR	Concept - Customer Lifetime Value	▼	>=	▼	50	▼
AND		▼	=	▼		▼
OR		▼	=	▼		▼

Fig. 7.19 Composed *alpha* condition

7.2.5 Results Evaluation

In order to look at the classification results, the fCQL fuzzy classification example and a fuzzy query where customers having a customer value higher than 50 are considered. As shown in Fig. 7.20, classification spaces with two dimensions are displayed in the query panel. Two major differences exist between the two-dimensional data analysis panel and the fuzzy classification panel. First the classification space is limited to the considered domains of the linguistic variables allowing a better visualization of the classified elements. Secondly, elements which do not satisfy the *with* condition are not considered into the statistics and the classification results but are also displayed (in a light gray color) in the space distribution in order to better evaluate the border between selected and not selected elements.

Similarly to the data analysis panel, the fuzzy classification information first gives the total number of elements as well as the number of selected elements. In this example, only 168 customers (16.8 %) have a customer value higher than 50 (see Fig. 7.21). Then for all the fuzzy classes the following information is available:

- *Number of elements*: The number of elements which have a positive membership degree in the given fuzzy class. All selected customers of this example partially belong to the fuzzy class C_1.
- *Number of fuzzy elements*: The number of fuzzy elements corresponds to the cardinality of the fuzzy class (see Definition 2.6) where the membership degrees of all the selected elements in the given class are summed. Since the number of elements previously computed is often not very useful, the number of fuzzy elements gives a much more precise information about the weight of the fuzzy class. In the fCQL example, the number of fuzzy elements in the fuzzy class C_1 is 55.26 which represents 32.89 % of the selected elements.
- *Class belonging*: The class belonging statistics tell the user about the minimum, maximum and average class belonging of the selected elements. In the example, the class belongings to the fuzzy class C_1 are 0.12, 0.75 and 0.33 respectively.

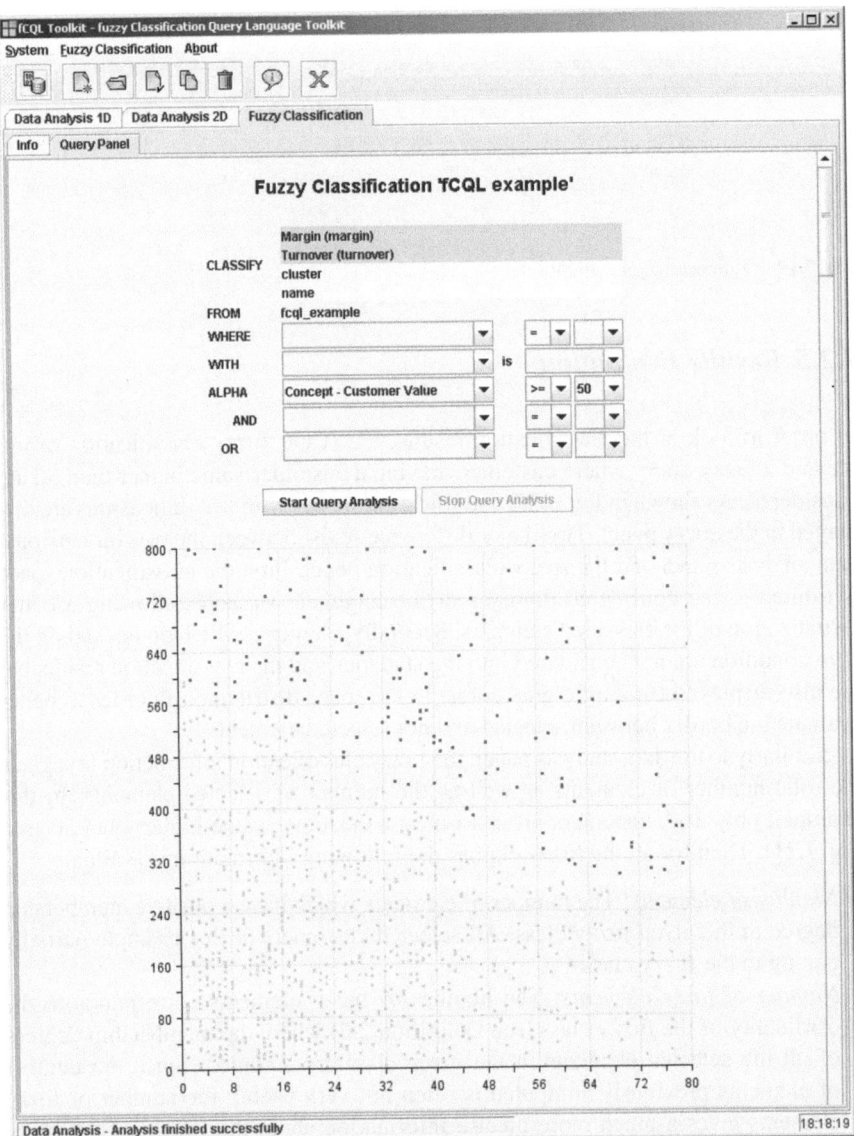

Fig. 7.20 Fuzzy classification query and data analysis space

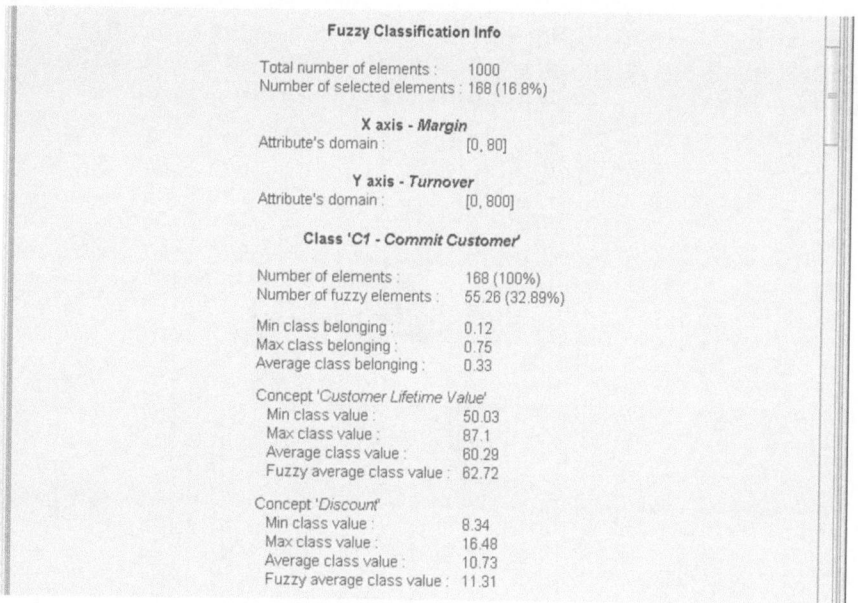

Fig. 7.21 Classification and class statistics

- *Concept statistics*: The statistics of all defined concepts are also available for each fuzzy class. These include the minimum, maximum, average and fuzzy average value of each concept. For instance, the concept customer lifetime value in the fuzzy class C_1 has the values 50.03, 87.1, 60.29 and 62.72 respectively.

Some statistical information as well as a distribution analysis are also available for each concept. As illustrated in Fig. 7.22, the minimum, the maximum and the average concept values are indicated. In the considered example these values are quite high. This is due to the fact that only customers having a customer lifetime value higher than 50 are taken into account.

Finally a subset of the fuzzy classification results is displayed in the results table. Since this table can contain a large number of elements, the displayed subset is limited to 100 elements and has only an illustration purpose. The results table contains for all the selected customers the values of the specified attributes and their membership degree in each fuzzy class. Two values of membership are given: the first, which is normally considered, is the normalized membership degree whereas the second (in parenthesis) is the raw membership degree. The normalization of the membership degrees is required when using the γ-operator for the aggregation. Finally, the values of all the defined concepts are given (i.e. the customer lifetime value and the discount values) (Fig. 7.23).

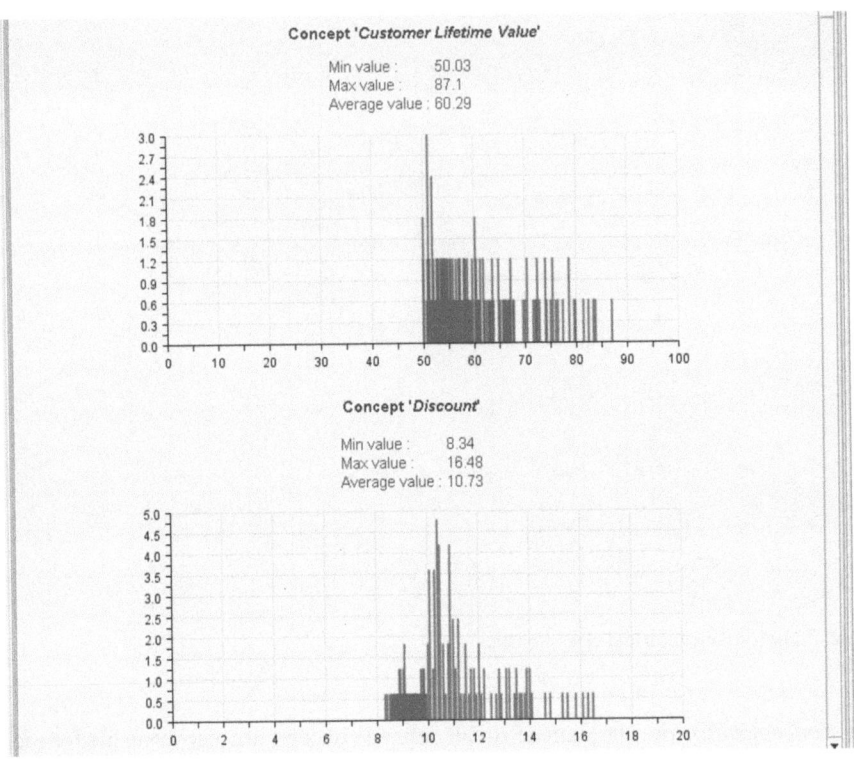

Fig. 7.22 Concepts' statistics

Id	Margin	Turnover	C1 - Commit Customer	C2 - Augment Margin	C3 - Augment Turnover	C4 - Don't Invest	Customer Lifetime Value	Discount
0	69	796	0.68 (0.93)	0.27 (0.37)	0.04 (0.06)	0.01 (0.01)	87.1	16.48
1	64	762	0.57 (0.87)	0.28 (0.43)	0.12 (0.18)	0.03 (0.05)	79.56	14.83
10	44	741	0.43 (0.7)	0.39 (0.63)	0.1 (0.15)	0.08 (0.13)	72.42	13.08
100	4	651	0.12 (0.18)	0.58 (0.88)	0.03 (0.05)	0.27 (0.41)	51.17	8.34
101	8	733	0.19 (0.29)	0.6 (0.9)	0.03 (0.04)	0.17 (0.26)	60.22	10.07
102	2	792	0.13 (0.18)	0.78 (0.98)	0 (0)	0.08 (0.1)	65.02	10.48
103	15	739	0.26 (0.4)	0.56 (0.86)	0.04 (0.06)	0.15 (0.23)	63.93	10.95
104	13	624	0.2 (0.32)	0.49 (0.79)	0.07 (0.11)	0.25 (0.4)	54.19	9.17
106	4	710	0.14 (0.2)	0.63 (0.92)	0.02 (0.03)	0.22 (0.32)	55.55	9.08
107	11	617	0.18 (0.29)	0.5 (0.8)	0.06 (0.1)	0.26 (0.42)	52.9	8.89
109	11	795	0.27 (0.37)	0.67 (0.93)	0.01 (0.01)	0.05 (0.07)	71.58	12.15

Fig. 7.23 Fuzzy classification results table

7.3 Database Schema of the Meta-Tables

This section describes the structure of the fCQL toolkit's meta-tables and their meaning. The relationships between the meta-tables are shown in Fig. 7.24. In order to set up a fuzzy classification, it is assumed that the data is available in a database

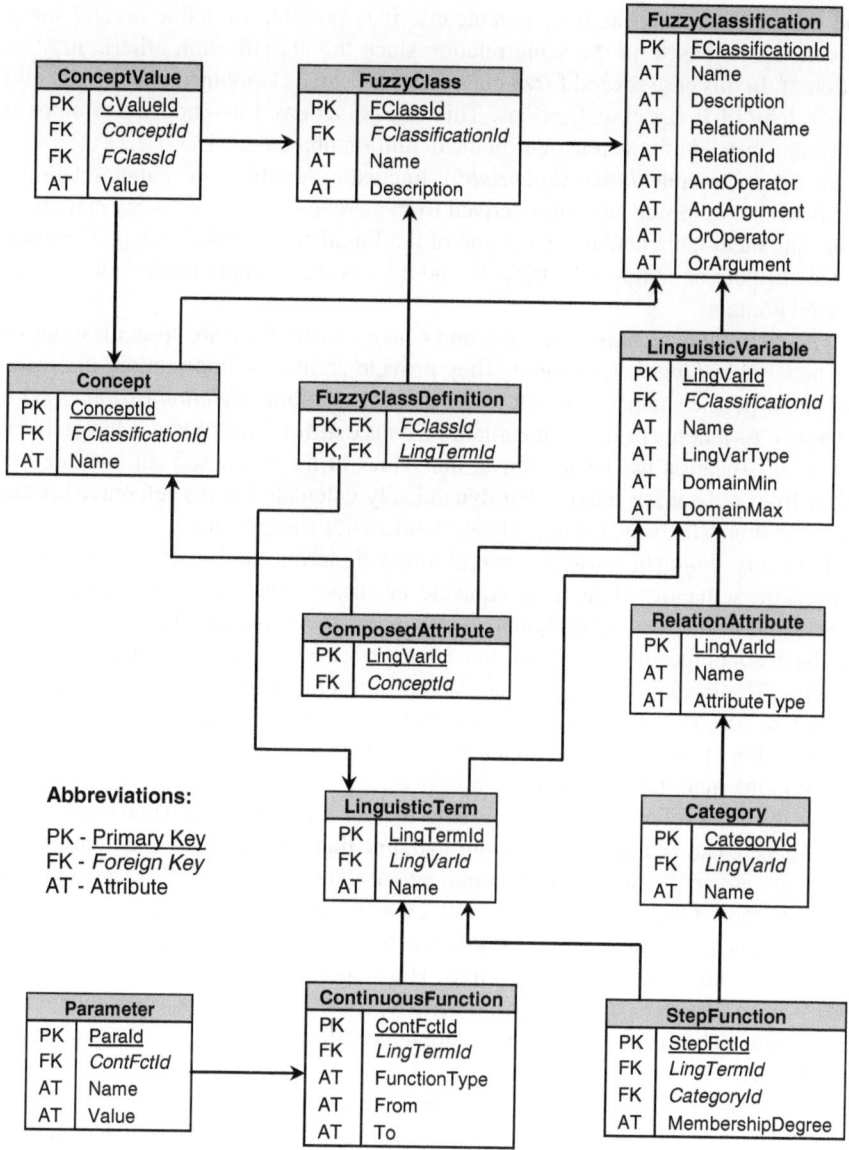

Fig. 7.24 Schema of the fCQL toolkit's meta-tables

relation or can be derived from different relations. This relation or view contains the elements to be classified as well as the qualifying attributes.

The meta-table *FuzzyClassification* defines fuzzy classifications by giving them a name and a description. It also specifies which relation holds the elements, which attribute identifies the elements and the aggregation operators for the intersection and the union as well as their parameters. It is possible to define several fuzzy classifications based on the same relation since the classification criteria may be different. In this case several fuzzy classifications can be combined in order to build a hierarchy of fuzzy classifications. This can be achieved by choosing composed attributes instead of concrete ones in the definition phase (see Sect. 7.2.3).

In the meta-table *LinguisticVariable*, linguistic variables are either relational attributes or composed attributes derived from a previously defined fuzzy classification. The meta-table contains the name of the linguistic variable, its type (i.e. relational attribute or composed attribute), and the lower and upper borders of the considered domain.

The meta-tables *RelationAttribute* and *ComposedAttribute* are specializations of the meta-table *LinguisticVariable*. They provide additional information, the name and the type (i.e. numeric or categoric) in the meta-table *RelationAttribute* and a reference to a concept in the meta-table *ComposedAttribute*. When a hierarchical fuzzy classification has been defined, the values of the composed attribute are not taken from a database relation but dynamically calculated in the referenced fuzzy classification. This way, the user always works with current data.

For every linguistic variable, several linguistic terms can be defined in order to express the semantics of the fuzzy equivalence classes. These terms are defined with their name in the meta-table *LinguisticTerm*. For the terms based on a categorical attribute, a membership degree has to be defined for every category of the attribute's domain. The available categories are stored in the meta-table *Category* which has a reference to a relational attribute and holds the name of the category. For each combination of categories and linguistic terms, the associated membership degree is stored in the meta-table *StepFunction*.

For the terms referring to a numeric attribute, several (continuous) functions, linear or S-shaped, can be composed in order to define their membership function. All the functions are implemented in the application code for efficiency reasons so that the meta-table *ContinuousFunction* only contains the function's type as well as its range. As the parameters of a linear function are different from those of an S-shaped function, another meta-table is required. The meta-table *Parameter* contains all the necessary parameters. In order to be generic, a parameter is described by a name and a value. This approach allows the implementation of new types of membership functions.

The meta-tables *Concept* and *ConceptValue* enable the user to assign to every fuzzy class one or more grades. These grades are necessary for the calculation of concepts which can be afterwards used in composed attributes. The meta-table *Concept* only contains a reference to the fuzzy classification and a name since the concept values are stored in the meta-table *ConceptValue*.

The fuzzy classes are defined in the *FuzzyClass* meta-table which contains their name and description. Finally the meta-table *FuzzyClassDefinition* associates the fuzzy classes with the linguistic terms of the linguistic variables.

Chapter 8
Conclusion

This concluding chapter first summarizes the key themes developed in this thesis in Sect. 8.1 and then discusses further developments as well as the wide variety of promising application domains in Sect. 8.2.

8.1 Summary

The fuzzy classification approach is an effective means for managing customer relationships. By providing accurate information about the classified elements, it allows companies to better achieve the CRM strategic objectives which are the acquisition, the retention and the recovery of customers.

As described in Chap. 5, fuzzy classes first enable an accurate customer positioning which results in efficient marketing campaigns through an adequate customer targeting and a precise controlling of the customer responses. It also allows marketers to implement trigger mechanisms to control the evolution of the customers in the classification space. Another important advantage offered by fuzzy classes is the personalization possibilities using the mass customization. As a result, personalized privileges or products, which are efficient means for the customer retention, can be automated. The mass customization approach can also be used to generate new concepts like the customer loyalty, which are vital indicators for companies. Furthermore the derived concepts can be used to build hierarchical fuzzy classifications (e.g. the customer lifetime value classification) which, by adequately integrating all the customers' information, lead to an accurate assessment of the customers. Controlling the customer relationships is also a much easier task with the help of a hierarchy of fuzzy classification where elements' properties can be analyzed through the different levels of the classification, each level containing semantic descriptions of the classes.

In Chap. 6, online customers have been analyzed in order to build an appropriate hierarchy of fuzzy classifications. Based on real life data, four representative customers of the Kiel & Co company have been studied at the different levels of the hierarchical fuzzy classification. The findings of this experiment are that the proposed

© Springer International Publishing Switzerland 2015
N. Werro, *Fuzzy Classification of Online Customers*, Fuzzy Management Methods,
DOI 10.1007/978-3-319-15970-6_8

hierarchical decomposition is valid and that the classification results derive new pertinent information about the customers.

The fCQL toolkit presented in Chap. 7 demonstrates that the fuzzy classification approach is technically realizable. The architecture of the fCQL toolkit allows the user to easily connect to any relational database system, to carry out basic data analysis, to define and to perform fuzzy classifications. The classification results are then returned with statistical information enabling a comprehensive analysis of the classified elements as well as of the fuzzy classification definition for optimization purpose. In order to concretely use the fCQL toolkit in the CRM field, different integration approaches of the fCQL toolkit within a CRM architecture have also been discussed in Chap. 4.

Besides the mentioned benefits, another key advantage derived from the fuzzy classification approach is its intuitive and human-oriented querying process. In Chap. 2, the affinity of fuzzy logic and human beings has been emphasized. With the use of linguistic variables and terms, human concepts can be first adequately modeled, then secondly, expressed on a linguistic level. The same applies to the fuzzy classes which are given a proper semantics. As a consequence, when querying a fuzzy classification with the fuzzy classification query language (see Chap. 3), the user does not have to deal with (sharp) numerical or categorical values but with predefined linguistic variables and terms and fuzzy classes whose meaning is linguistically expressed. With the extension of the fuzzy Classification Query Language by the new *alpha* clause, the user is also able to precisely target customer groups based on class belonging information as well as on derived concepts.

From the company's perspective, linguistic variables and terms also offer a way of defining a common terminology across all departments, i.e. engineers as well as managers can use the same vocabulary. Another important aspect is that the definition of fuzzy classifications should only be carried out by experts of the domain; using predefined linguistic variables and terms hides the complexity of the domain from the end users and also enables optimization of the fuzzy classification definition without any impact on the end users.

Finally it should be reminded that the fuzzy classification does not require any transformation nor any migration of the business data (potentially very large in volume) since all the fuzzy classification definitions are stored separately in meta-tables.

Using a fuzzy classification however does also have drawbacks: the main issue concerns the modeling of the fuzzy classifications since more parameters have to be determined. In comparison with a sharp classification which requires the definition of qualifying attributes and equivalence classes, the fuzzy classification requires the additional definition steps regarding the membership functions as well as the determination of the aggregation operators and their arguments (if any). If a hierarchical fuzzy classification is considered, concept values have to be associated with the fuzzy classes and a decomposition based on the available attributes has to be worked out. The fuzzy classification definition process is therefore more complex than for traditional classification approaches. As a consequence, the participation of the management, marketing specialists and database architects is even more important for the modeling of a hierarchical fuzzy classification.

8.2 Outlook

On an application level, the proposed fuzzy classification approach opens the door to many application fields. Even though it can be adopted in almost all application domains where data analysis comes into play (e.g. segmentation, data complexity reduction, etc.), the following managerial application domains seem most promising:

- *Marketing Mix Theory*: The marketing mix theory with its four P's (Product, Price, Place and Promotion) positions products on the marketplace. Since Sect. 5.3.2 introduced the notion of fuzzy product portfolio and a fuzzy price can be easily calculated as a concept (or determined using a personalized discount), a fuzzy marketing mix theory could be achieved by 'fuzzifying' the place and promotion components, especially for e-marketing.
- *Portfolio Management*: Portfolios are typical examples of sharp classifications. Considering for instance the SWOT (Strengths, Weaknesses, Opportunities and Threats) analysis, elements located on the borders are sharply classified as either a strength or a threat. Clearly, portfolios with fuzzy classes would be much more realistic and provide better information for decision making [141, 142].
- *Performance Measurement and Customer Performance Measurement*: Performance measurement relies on many key performance indicators (both quantitative and qualitative) [141]. A hierarchical fuzzy classification with concepts can derive the desired performance measures and enables the control of customer performance within the classification hierarchy.
- *Risk Management*: In banking or insurance, individuals or companies have to be divided into risk classes. Very often, pricing components directly depend on risk levels. With a fuzzy classification, the calculation of risk degrees, creditworthiness or other indicators can be carried out with finer granularity.
- *Strategic Management*: For the analysis of markets, fuzzy classification allows demographic, geographic, behavioral and psychographic market segmentations. It is more successful and realistic to fuzzily target markets and to fuzzily position brands or companies in their markets.
- *Supply Chain Management*: With a fuzzy approach, it is possible to classify, analyze and evaluate different suppliers and their delivery processes. A fuzzy supplier rating and a fuzzy judgment of quality and time schedules of the delivery processes provide for more differentiated planning. For instance, improvements in the delivery system can be effected by observing moving targets in fuzzy classes.
- *Total Quality Management*: Quality measures are not only numeric; there are also qualitative measures. The equal treatment of quantitative and qualitative properties makes the fuzzy classification approach attractive for TQM. It is possible to fuzzily categorize, analyze and control materials, products, services and processes.

On a conceptual level, considering a given application domain, further general or specialized (i.e. context specific) indicators derived from the fuzzy classification information could be discovered, extending this way the available fuzzy classification toolkit. For instance, Zumstein [141] developed a balance indicator in the field of portfolio management.

On an implementation level, many improvements can be achieved. Beside minor and context related enhancements, the fCQL toolkit could be extended in the following directions:

- The ability to connect a data warehouse would be most useful for data analysis in large enterprises since elements could be analyzed over time and at the different aggregation levels.
- A next step is the development of fuzzy data warehouses since introducing equivalence classes of attributes produces a multi-dimensional fuzzy data cube. For this purpose, the main operators of a data warehouse, i.e. drill-down, roll-up, slicing and dicing, have to be extended.
- The fCQL toolkit is actually limited to the graphical representation of two-dimensional classification spaces. The visualization of multi-dimensional spaces would be an appreciable add-on.
- In order to facilitate the fuzzy classification definition process, further data analysis panels using data mining techniques (e.g. classification and clustering methods) could be integrated. Also, a design framework for fuzzy classification should be developed.
- In order to enable the analysis of incomplete data, the handling of missing values could be implemented. Actually elements with missing values are not taken into account.
- Some import and export interfaces could be implemented in order to be able to archive or exchange classification results. The querying of fuzzy classification could also be offered as a web service so that the fCQL toolkit could be queried online or easily integrated within a SOA architecture.

Appendix A
Fuzzy Set Operators

This appendix gives non-exhaustive lists of t-norm (Table A.1), t-conorm (Table A.2), averaging (Table A.3) and compensatory (Table A.4) operators, as well as linguistic modifiers (Table A.5), commonly found in the literature.

Table A.1 Examples of t-norm operators

Operator	Expression
Minimum (t_1)	$f(x, y) = min(x, y)$
Hamacher product (t_2)	$f(x, y) = \frac{xy}{p+(1-p)(x+y-xy)}$, $p_2 \geq 0$
Algebraic product (t_3)	$f(x, y) = xy$
Einstein product (t_4)	$f(x, y) = \frac{xy}{2-(x+y-xy)}$
Bounded product (t_5)	$f(x, y) = max[0, (1 + p)(x + y - 1) - pxy)]$, $p_5 \geq -1$
Drastic product (t_6)	$f(x, y) = \begin{cases} min(x, y) & \text{if } max(x, y) = 1 \\ 0 & \text{otherwise} \end{cases}$
Yager family	$f(x, y) = 1 - min(1, [(1 - x)^p + (1 - y)^p]^{\frac{1}{p}}$, $p \geq 1$
Sugeno family	$f(x, y) = min(1, x + y + pxy)$, $p \geq -1$
Dubois-Prade family	$f(x, y) = \frac{xy}{max(x,y,p)}$, $p \in [0, 1]$
Frank family	$f(x, y) = log_p \left(1 + \frac{(p^x-1)(p^y-1)}{p-1}\right)$, $p > 0$ and $p \neq 1$
Others	$f(x, y) = \frac{1}{1+\left[\left(\frac{1-x}{x}\right)^p+\left(\frac{1-y}{y}\right)^p\right]^{\frac{1}{p}}}$, $p > 0$
	$f(x, y) = \frac{1}{(\frac{1}{x^p} + \frac{1}{y^p} - 1)}$
	$f(x, y) = [max(0, x^p + y^p - 1)]^{\frac{1}{p}}$

where $t_1 \geq t_2 \geq t_3 \geq t_4 \geq t_5 \geq t_6$ for p_2, $p_5 = 0$ [30].

© Springer International Publishing Switzerland 2015
N. Werro, *Fuzzy Classification of Online Customers*, Fuzzy Management Methods,
DOI 10.1007/978-3-319-15970-6

Table A.2 Examples of t-conorm operators

Operator	Expression
Maximum (s_1)	$f(x, y) = max(x, y)$
Hamacher sum (s_2)	$f(x, y) = \frac{x+y+(p-1)xy}{1+pxy}$, $p_2 \geq -1$
Algebraic sum (s_3)	$f(x, y) = x + y - xy$
Einstein sum (s_4)	$f(x, y) = \frac{x+y}{1+xy}$
Bounded sum (s_5)	$f(x, y) = min(1, x + y + pxy)$, $p_5 \geq 0$
Drastic sum (s_6)	$f(x, y) = \begin{cases} max(x, y) & \text{if } min(x, y) = 0 \\ 1 & \text{otherwise} \end{cases}$
Yager family	$f(x, y) = min(1, [x^p + y^p]^{\frac{1}{p}})$, $p \geq 1$
Sugeno family	$f(x, y) = min(1, x + y + p - xy)$, $p \geq 0$
Dubois-Prade family	$f(x, y) = \frac{x+y-xy-min(x,y,(1-p))}{max(1-x,1-y,p)}$, $p \in [0, 1]$
Frank family	$f(x, y) = log_p \left(1 + \frac{(p^{1-x}-1)(p^{1-y}-1)}{p-1}\right)$, $p > 0$ and $p \neq 1$
Others	$f(x, y) = \dfrac{1}{1-\left[\frac{x}{(1-x)^p}+\frac{y}{(1-y)^p}\right]^{\frac{1}{p}}}$, $p > 0$
	$f(x, y) = \dfrac{1}{1-\left[\frac{1}{(1-x)^p}+\frac{1}{(1-y)^p}-1\right]^{\frac{1}{p}}}$, $p > 0$
	$f(x, y) = 1 - max(0, [(1 - x)^p + (1 - y)^p - 1]^{\frac{1}{p}})$, $p > 0$
	$f(x, y) = \frac{x+y-xy-(1-p)xy}{1-(1-p)xy}$, $p \geq 0$

where $s_1 \leq s_2 \leq s_3 \leq s_4 \leq s_5 \leq s_6$ for $p_2 = -1$ and $p_5 = 0$ [30].

Table A.3 Examples of averaging operators

Operator	Expression
Fuzzy and	$f(x, y) = \gamma \, min(x, y) + \frac{(1-\gamma)}{2}(x + y)$, $\gamma \in [0, 1]$
Fuzzy or	$f(x, y) = \gamma \, max(x, y) + \frac{(1-\gamma)}{2}(x + y)$, $\gamma \in [0, 1]$
Others	$f(x, y) = \gamma \, min(x, y) + (1 - \gamma) \, max(x, y)$, $\gamma \in [0, 1]$
	$f(x, y) = \frac{x+y-xy}{1+x+y-2xy}$
	$f(x, y) = \frac{xy}{1+x-y+2xy}$
	$f(x, y) = \frac{max(x,y)}{1+|x-y|}$
	$f(x, y) = \frac{min(x,y)}{1+|x-y|}$

Table A.4 Examples of compensatory operators

Operator	Expression
γ-operator	$f(x_i) = \left(\prod_{i=1}^{m} x_i\right)^{(1-\gamma)} \left(1 - \prod_{i=1}^{m}(1 - x_i)\right)^{\gamma}$, $\gamma \in [0, 1]$
Others	$f(x, y) = \gamma \, xy + (1 - \gamma)(x + y - xy)$, $\gamma \in [0, 1]$

Table A.5 Examples of linguistic modifiers

Linguistic expression	Operator	Expression
Very	Concentration	$f(x) = x^2$
More or less	Dilation	$f(x) = x^{\frac{1}{2}}$
Extremely	Concentration	$f(x) = x^3$
Slightly	Dilation	$f(x) = x^{\frac{1}{3}}$

Appendix B
Query Languages' Grammar

B.1 Original fCQL Grammar

<ClassificationQuery>=
  classifiy <AttributeList>
  from <RelationName>
  { **with** <ClassificationCondition> }

<AttributeList>=
  ColumnDefinition { , <AttributeList> }

<RelationName>=
  RelationIdentifier | *ViewIdentifier*

<ClassificationCondition>=
  <ClassSelection>{ or <ClassificationCondition> } |
  <LinguisticVariableSelection> { or <ClassificationCondition> }

<ClassSelection>=
  class is *ClassDefinition*

<LinguisticVariableSelection>=
  LinguisticVariableDefinition **is** <TermCondition>
  { **and** <LinguisticVariableSelection> }

<TermCondition>=
  TermDefinition { **or** <TermCondition> }

© Springer International Publishing Switzerland 2015
N. Werro, *Fuzzy Classification of Online Customers*, Fuzzy Management Methods,
DOI 10.1007/978-3-319-15970-6

B.2 FQUERY Grammar

```
<Query>=
    select <ListOfFields>
    from <ListOfTables>
    where <Condition>

<Condition>=
    <LinguisticQuantifier> <SequenceOfSubconditions>

<SequenceOfSubconditions>=
    <Subcondition> |
    <Subcondition> or <SequenceOfSubconditions>

<Subcondition>=
    <LinguisticQuantifier> <ImportanceCoefficient>
    <SequenceOfAtomicConditions>

<SequenceOfAtomicConditions>=
    <AtomicCondition> |
    <AtomicCondition> and <SequenceOfAtomicConditions>

<AtomicCondition>=
    <Attribute> = <FuzzyValue> |
    <Attribute> <RelationalOperator> <NumericalAttribute> |
    <Attribute> <FuzzyRelation> <Attribute> |
    <Attribute> <FuzzyRelation> <Number> |
    <SingleValuedAttribute> in <FuzzySetConstant> |
    <MultiValuedAttribute> <CompatibilityOperator>
    <FuzzySetConstant>

<Attribute>=
    <NumericField>

<LinguisticQuantifier>=
    <OWA-Tag> <QuantifierName>

<OWA-Tag>=
    OWA |

<RelationalOperator>=
    < | <= | > | >= | =
```

References

1. C. Alsina, On a family of connectives for fuzzy sets. Fuzzy Sets Syst. **16**, 231–235 (1985)
2. Aristotle, Aristotle's metaphysics. Trans. H. G. Apostle (Indiana University Press, Bloomington, 1966)
3. D. Barbará, H. Garcia-Molina, D. Porter, The management of probabilistic data. IEEE Trans. Knowl. Data Eng. **4**(5), 487–502 (1992)
4. A. Bauer, H. Günzel, *Data Warehouse Systeme*, 2nd edn. (dpunkt Publisher, Heidelberg, 2004)
5. M.J.A. Berry, G.S. Linoff, *Data Mining Techniques for Marketing, Sales and Cutomer Support* (Wiley, New York, 1997)
6. M.J.A. Berry, G.S. Linoff, *Mastering Data Mining* (Wiley, New York, 2000)
7. A. Berson, S. Smith, K. Thearling, *Building Data Mining Applications for CRM* (McGraw-Hill, New York, 2000)
8. R.C. Blattberg, G. Getz, J.S. Thomas, *Customer Equity—Building and Managing Relationships as Valuable Assets* (Harvard Business School Press, Boston, 2001)
9. P. Bosc, M. Galibourg, G. Hamon, Fuzzy querying with SQL: extensions and implementation aspects. Fuzzy Sets Syst. **28**, 333–349 (1988)
10. P. Bosc, L. Lietard, O. Pivert, Quantified statements in a flexible relational query language. in *Proceedings of the ACM Symposium on Applied Computing* (Nashville, 1995), pp. 488–492
11. P. Bosc, O. Pivert, Fuzzy querying in conventional databases, in *Fuzzy Logic for the Management of Uncertainty*, ed. by L.A. Zadeh, J. Kacprzyk (Wiley, New York, 1992), pp. 645–671
12. P. Bosc, O. Pivert, Fuzzy queries and relational databases. in *Proceedings of the ACM Symposium on Applied Computing* (Phoenix, 1994) pp. 170–174
13. P. Bosc, O. Pivert, SQLf: a relational database language for fuzzy querying. IEEE Trans. Fuzzy Syst. **3**(1), 1–17 (1995)
14. P. Bosc, O. Pivert, SQLf query functionality on top of a regular relational database management system, in *Knowledge Management in Fuzzy Databases, vol. 39, Studies in Fuzziness and Soft Computing*, ed. by O. Pons, M.A. Vila, J. Kacprzyk (Physica Publisher, Heidelberg, 2000), pp. 171–190
15. M. Bruhn, *Integrierte Kundenorientierung* (Gabler Publisher, Wiesbaden, 2002)
16. M. Bruhn, *Relationship Marketing—Management of Customer Relationships* (Prentice Hall, Harlow, 2003)
17. M. Bruhn, C. Homburg, *Gabler Marketing Lexikon* (Gabler Publisher, Wiesbaden, 2001)
18. R. Cavallo, M. Pittarelli, The theory of probabilistic databases. in *Proceedings of the 13th International Conference on Very Large Data Bases, VLDB '87* (1987) pp. 71–81
19. D.D. Chamberlin, M.M. Astrahan, K.P. Eswaran, P.P. Griffiths, R.A. Lorie, J.W. Mehl, P. Reisner, B.W. Wade, A unified approach to data definition, manipulation and control. IBM J. Res. Dev. **20**(6), 560–575 (1976)

© Springer International Publishing Switzerland 2015

N. Werro, *Fuzzy Classification of Online Customers*, Fuzzy Management Methods,
DOI 10.1007/978-3-319-15970-6

20. G. Chen, *Fuzzy Logic in Data Modeling—Semantics, Constraints and Database Design* (Kluwer Academic Publishers, London, 1998)

21. P.P. Chen, The entity-relationship model toward a unified view of data. ACM Trans. Datab. Syst. **1**(1), 9–36 (1976)

22. E.F. Codd, A relational model of data for large shared data banks. Comm. ACM **13**(6), 377–387 (1970)

23. E.F. Codd, Extending the database relational model to capture more meaning. ACM Trans. Datab. Syst. **4**(4), 397–434 (1979)

24. E.F. Codd, Missing information (applicable and inapplicable) in relational databases. ACM SIGMOD Rec. **15**(4), 53–78 (1986)

25. E.F. Codd, More commentary on missing information in relational databases (applicable and inapplicable information). ACM SIGMOD Rec. **16**(1), 42–50 (1987)

26. E. Cox, *The Fuzzy Systems Handbook*, 2nd edn. (Academic Press, San Diego, 1999)

27. D. Diller, Die Bedeutung des Beziehungsmarketing für den Unternehmenserfolg, in *Grundlagen des CRM—Konzepte und Gestaltung*, 2nd edn. ed. by H. Hippner, K.D. Wilde (Gabler Publisher, Wiesbaden, 2006), pp. 97–120

28. J. Dombi, Membership function as an evaluation. Fuzzy Sets Syst. **35**(1), 1–21 (1990)

29. D. Driankov, H. Hellendoorn, M. Reinfrank, *An Introduction to Fuzzy Control* (Springer, Berlin, 1993)

30. D. Dubois, H. Prade, A class of fuzzy measures based on triangular norms. Int. J. General Syst. **8**, 43–61 (1982)

31. D. Dubois, H. Prade, Criteria aggregation and ranking of alternatives in the framework of fuzzy set theory, in *Fuzzy Sets and Decision Analysis*, ed. by H.-J. Zimmermann, L.A. Zadeh, B.R. Gaines (North-Holland, Amsterdam, 1984), pp. 209–240

32. e-Business W@tch. The European e-Business Report 2006/2007 edition—a portrait of e-Business in 10 sectors of the EU Economy, http://www.ebusiness-watch.org. Accessed 26 Sept 2007

33. A. Edmunds, A. Morris, The problem of information overload in business organisations: a review of the literature. Int. J. Inf. Manage. **20**(1), 17–28 (2000)

34. R. Elmasri, S.B. Navathe, *Fundamentals of Database Systems*, 3rd edn. (Addison-Wesley, New York, 2000)

35. S. Finnerty, S. Shenoi, Abstraction-based query languages for relational databases, in *Advances in Fuzzy Theory and Technology*, vol. 1, ed. by P. Wang (Bookwrights, Durham, 1993), pp. 195–218

36. D. Flanagan, *Java in a Nutshell*, 5th edn. (O'Reilly, Sebastopol, 2005)

37. D. Frauchiger, A. Meier, H. Stormer, N. Werro, Zur Entwicklung des Struts-basierten Webshops eSarine. HMD—Praxis der Wirtschaftsinformatik **238**, 62–71 (2004)

38. R. Fullér, H.-J. Zimmermann, On Zadeh's compositional rule of inference, in *Fuzzy Logic, vol. 12, System Theory, Knowledge Engineering and Problem Solving*, ed. by R. Lowen, M. Roubens (Kluwer Academic, Dordrecht, 1993), pp. 193–200

39. J. Galindo, J.M. Medina, M.C. Aranda, Querying fuzzy relational databases through fuzzy domain calculus. Int. J. Intell. Syst. **14**, 375–411 (1999)

40. J. Galindo, J.M. Medina, O. Pons, J.C. Cubero, A server for fuzzy SQL queries. in *Flexible Query Answering Systems, volume 1495 of Lecture Notes in Artificial Intelligence* ed by. T. Andreasen, H. Christiansen, H.L. Larsen (Springer, Berlin, 1998)

41. J. Galindo, A. Urrutia, M. Piattini, *Fuzzy Databases—Modeling, Design and Implementation* (Idea Group Publishing, Hershey, 2006)

42. T. Gawlik, J. Kellner, D. Seifert, *Effiziente Kundenbindung mit CRM* (Galileo Press, Bonn, 2002)

43. J. Godjevac, *Neuro-Fuzzy Controllers—Design and Application* (Presses Polytechniques et Universitaires Romandes, Lausanne, 1997)

44. J. Godjevac, *Idées Nettes sur la Logique Floue* (Presses Polytechniques et Universitaires Romandes, Lausanne, 1999)

45. J. Grant, Incomplete Information in a Relational Database. Fundamenta Informaticae **3**(3), 363–378 (1980)
46. M.M. Gupta, Fuzzy logic and neural systems. in *International Series in Intelligent Technologies* ed. by D. Ruan (Kluwer Academic, 1995), pp 225–244
47. J. Hale, S. Shenoi, Catalytic inference analysis using fuzzy relations. in *Proceedings of the 6th International Fuzzy Systems Association World Congress, IFSA 1995*, vol. 2 (Sao Paulo, Brazil, 1995), pp. 413–416
48. T. Harrison, *Financial Services Marketing* (Pearson Education, Essex, 2000)
49. H. Hellendoorn, Fuzzy control: an overview, in *Fuzzy Systems in Computer Science*, ed. by R. Kruse, J. Gebhardt, R. Palm (Vieweg, Wiesbaden, 1994), pp. 11–27
50. A.R. Hevner, S.T. March, J. Park, S. Ram, Design science in information systems research. MIS Quarterly **28**(1), 75–105 (2004)
51. H. Hippner, CRM—Grundlagen, Ziele und Konzepte, in *Grundlagen des CRM—Konzepte und Gestaltung*, 2nd edn., ed. by H. Hippner, K.D. Wilde (Gabler Publisher, Wiesbaden, 2006), pp. 17–44
52. H. Hippner, R. Rentzmann, K.D. Wilde, Aufbau und Funktionalitäten von CRM-Systemen, in *Grundlagen des CRM—Konzepte und Gestaltung*, 2nd edn. ed. by H. Hippner, K.D. Wilde (Gabler Publisher, Wiesbaden, 2006), pp. 45–74
53. H. Hippner, K.D. Wilde, CRM—Ein Ueberblick, in *Effektives Customer Relationship Management*, 2nd edn. ed. by S. Helmke, W. Dangelmaier (Gabler Publisher, Wiesbaden, 2002), pp. 3–37
54. H. Hippner, K.D. Wilde, Data mining im CRM, in *Effektives Customer Relationship Management*, 2nd edn. ed. by S. Helmke, W. Dangelmaier (Gabler Publisher, Wiesbaden, 2002), pp. 211–231
55. H. Hippner, K.D. Wilde, Informationstechnologische Grundlagen der Kundenbindung, in *Handbuch Kundenbindungsmanagement*, 4th edn. ed. by M. Bruhn, C. Homburg (Gabler Publisher, Wiesbaden, 2003), pp. 451–481
56. C. Homburg, M. Bruhn, Kundenbindungsmanagement—Eine Einführung in die theoretischen und praktischen Problemstellungen, in *Handbuch Kundenbindungsmanagement*, 4th edn. ed. by M. Bruhn, C. Homburg (Gabler Publisher, Wiesbaden, 2003), pp. 3–37
57. C. Homburg, M. Fassnacht, Kundennähe, Kundenzufriedenheit und Kundenbindung bei Dienstleistungsunternehmen. in *Handbuch Dienstleistungsmanagement*, ed. by M. Bruhn and H. Meffert (Gabler Publisher, 1998), pp. 405–428
58. W.H. Inmon, *Building the Data Warehouse*, 2nd edn. (Wiley, New York, 1996)
59. K. Inoue, H. Nakajima, N. Yoshikawa, Pricing strategies in the E-business age. Nomura Res. Inst. Pap. **23**, 1–9 (2001)
60. J. Kacprzyk, Fuzzy logic in DBMSs and querying. in *Proceedings of the 2nd New Zealand Two-Stream International Conference on Artificial Neural Networks and Expert Systems* (Dunedin, 1995), pp. 106–109
61. J. Kacprzyk, S. Zadrozny, FQUERY for access: fuzzy querying for a windows-based DBMS, in *Fuzziness in Database Management Systems, vol. 5, Studies in Fuzziness*, ed. by P. Bosc, J. Kacprzyk (Physica Publisher, Heidelberg, 1995), pp. 415–433
62. J. Kacprzyk, S. Zadrozny, Multi-valued fields and values in fuzzy querying via FQUERY for access. in *Proceedings of the Fifth IEEE International Conference on Fuzzy Systems*, vol. 2, (New Orleans, 1996), pp. 1351–1357
63. J. Kacprzyk, S. Zadrozny, Data mining via fuzzy querying over the internet, in *Knowledge Management in Fuzzy Databases, vol. 39, Studies in Fuzziness and Soft Computing*, ed. by O. Pons, M.A. Vila, J. Kacprzyk (Physica Publisher, Heidelberg, 2000), pp. 211–233
64. J. Kacprzyk, S. Zadrozny, SQLf and FQUERY for access. in *Proceedings of the International Fuzzy Systems Association World Congress and the North American Fuzzy Information Processing Society, IFSA/NAFIPS 2001*, vol. 4 (Vancouver, 2001), pp. 2464–2469
65. N.K. Kasabor, *Foundations of Neural Networks, Fuzzy Systems, and Knowledge Engineering* (MIT Press, Cambridge, 1996)
66. E.P. Klement, R. Mesiar, E. Pap, *Triangular Norms* (Kluwer Academic, Dordrecht, 2000)

67. P. Kotler, C.J. Dipak, S. Maesincee, *Marketing Moves—A New Approach to Profits, Growth, and Renewal* (Harward Business School Press, Boston, 2002)
68. R. Kruse, J. Gebhardt, F. Klawonn, *Foundations of Fuzzy Systems* (Wiley, Chichester, 1994)
69. E.S. Lee, Q. Zhu, *Fuzzy and Evidence Reasoning, vol. 6. Studies in Fuzziness* (Physica-Verlag, Heidelberg, 1995)
70. R. Lefébure, G. Venturi, *Gestion de la Relation Client* (Editions Eyrolles, Paris, 2000)
71. G. Linden, B. Smith, J. York, Amazon.com recommendations: item-to-item collaborative filtering. IEEE Internet Comput. **7**(1), 76–80 (2003)
72. D. Maier, *The Theory of Relational Databases* (Computer Science Press, Rockville, 1983)
73. U. Manber, A. Patel, J. Robison, The business of personalization—experience with personalization of Yahoo!. Comm. ACM **43**(8), 35–39 (2000)
74. S.T. March, G.F. Smith, Design and natural science research on information technology. Decis. Support Syst. **15**, 251–266 (1995)
75. A. Meier, Vom Digitalen Produkt bis hin zum customer relationship management, in *Internet & Electronic Business—Herausforderung an das Management*, ed. by A. Meier (Orell Füssli Publisher, Zürich, 2001), pp. 11–26
76. A. Meier, Organisation, Implementierung und Controlling des Kundenbeziehungsmanagements. Die Unternehmung **5**, 331–346 (2004)
77. A. Meier, *Relationale und postrelationale Datenbanken*, 6th edn. (Springer, Heidelberg, 2007)
78. A. Meier, C. Mezger, N. Werro, G. Schindler, Zur Unscharfen Klassifikation von Datenbanken mit fCQL, *Proceedings of the GI-Workshop LLWA–Lehren, Lernen, Wissen, Adaptivität* (Karlsruhe, Germany, October 2003), pp. 151–158
79. A. Meier, C. Savary, G. Schindler, Y. Veryha, Database schema with fuzzy classification and classification query language. in *Proceedings of the International Congress on Computational Intelligence—Methods and Applications, CIMA 2001* (Bangor, UK, June 2001), pp. 1–7
80. A. Meier, G. Schindler, N. Werro, Fuzzy classification on relational databases. in *Handbook of Research on Fuzzy Information Processing in Databases*, ed. by J. Galindo (Idea Group Publishing, Hershey, in press, 2008)
81. A. Meier, H. Stormer, *eBusiness & eCommerce* (Springer Publisher, Heidelberg, 2005)
82. A. Meier, N. Werro, Extending a webshop with a fuzzy classification model for online customers. in *Proceedings of the International Association for Development of the Information Society Conference*, vol. 1. IADIS e-Society 2006 (Dublin, Ireland, 2006), pp. 305–312
83. A. Meier, N. Werro, A fuzzy classification model for online customers. Informatica Int. J. Comput. Inform. **31**, 175–182 (2007)
84. A. Meier, N. Werro, M. Albrecht, M. Sarakinos, Using a Fuzzy Classification Query Language for Customer Relationship Management. in *Proceedings of the 31st International Conference on Very Large Data Bases, VLDB 2005* (Trondheim, Norway, 2005), pp. 1089–1096
85. K.D. Meyer-Gramann, Fuzzy classification: an overview, in *Fuzzy Systems in Computer Science*, ed. by R. Kruse, J. Gebhardt, R. Palm (Vieweg, Wiesbaden, 1994), pp. 277–294
86. W.W. Moe, P.S. Fader, Capturing evolving visit behaviour in clickstream data. J. Interact. Market. **18**(1), 5–19 (2004)
87. N. Mouaddib, Fuzzy identification in fuzzy databases—the Nuanced relation division. Int. J. Intell. Syst. **9**, 461–473 (1994)
88. M. Mukaidono, *Fuzzy Logic for Beginners* (World Scientific Publishing, London, 2001)
89. C. Nançoz, *mEdit—Membership Function Editor for fCQL-based Architecture*. Master's thesis (Department of Informatics, University of Fribourg, Switzerland, 2004)
90. P. Neckel, B. Knobloch, *Customer Relationship Analytics* (dpunkt Publisher, Heidelberg, 2005)
91. A. Newell, H.A. Simon, *Human Problem Solving* (Prentice Hall, Englewood Cliffs, New Jersey, 1972)
92. H.T. Nguyen, E.A. Walker, *A First Course in Fuzzy Logic* (CRC Press, Boca Raton, 1997)
93. P.T. Nguyen, G. Cliquet, A. Borges, F. Leray, L'opposition entre Taille du Marché et Degré d'Homogénéité des Segments: une Approche par la Logique Floue. Décis. Market. **32**, 55–69 (2003)

94. V. Novák, Fuzzy sets in natural language processing, in *An Introduction to Fuzzy Logic Applications in Intelligent Systems*, ed. by R.R. Yager, L.A. Zadeh (Kluwer Academic, Dordrecht, 1992), pp. 185–200

95. J. Paredaens, P. De Bra, M. Gyssens, D. Van Gucht, *The Structure of the Relational Database Model* (Springer, New York, 1989)

96. W. Pedrycz, F. Gomide, *An Introduction to Fuzzy Sets—Analysis and Design* (MIT Press, Cambridge, 1998)

97. B.J. Pine, S. Davis, *Mass Customization—The New Frontier in Business Competition* (Harvard Business School Press, Boston, 1999)

98. H. Rheingold, *The Virtual Community—Homesteading on the Electronic Frontier* (Addison Wesley, New York, 1993)

99. D. Risch, P. Schubert, Customer profiles, personalization and privacy. in *Proceedings of CollECTeR Europe 2005* (Furtwangen, Germany, 2005), pp. 1–12

100. R.T. Rust, V.A. Zeithaml, K.N. Lemon, *Driving Customer Equity* (Free Press, New York, 2000)

101. C. Schaller, C.M. Stotko, F.T. Piller, Mit Mass Customization basiertem CRM zu Loyalen Kundenbeziehungen, in *Grundlagen des CRM—Konzepte und Gestaltung*, 2nd edn., ed. by H. Hippner, K.D. Wilde (Gabler Publisher, Wiesbaden, 2006), pp. 121–143

102. G. Schindler, *Fuzzy Datenanalyse durch Kontexbasierte Datenbankanfragen* (Deutscher Universitäts Verlag, Wiesbaden, 1998)

103. P. Schubert, *Virtuelle Transaktionsgemeinschaften im Electronic Commerce* (Josef Eul Publisher, Lohmar, 1999)

104. P. Schubert, Einführung in die E-Business-Begriffswelt, in *E-Business erfolgreich planen und realisieren—case studies von zukunftsorientierten Unternehmen*, ed. by P. Schubert, R. Wölfle (Hanser Publisher, München, 2000), pp. 1–12

105. P. Schubert, The pivotal role of community building in electronic commerce. in *Proceedings of the 33th Hawaii International Conference on System Sciences, HICSS 2000*, vol. 1. (Hawaii, 2000)

106. P. Schubert, Personalizing E-Commerce applications in SMEs. in *Proceedings of the Ninth Americas Conference on Information Systems, AMCIS* (Tampa, Florida, USA, 2003), pp. 737–750

107. P. Schubert, R. Wölfle (eds.), *E-Business erfolgreich planen und realisieren—case studies von zukunftsorientierten Unternehmen* (Hanser Publisher, München, 2000)

108. P. Schubert, R. Wölfle, W. Dettling (eds.), *Fulfillment im E-Business—Praxiskonzepte innovativer Unternehmen* (Hanser Publisher, München, 2001)

109. P. Schubert, R. Wölfle, W. Dettling (eds.), *Procurement im E-Business—Einkaufs und Verkaufsprozesse elektronisch optimieren* (Hanser Publisher, München, 2002)

110. C.E. Shannon, A mathematical theory of communication. Bell syst. Tech. J. 27:379–423, 623–656 (1948)

111. S. Shenoi, On classificalizing fuzzy databases. in *Proceedings of the 5th International Fuzzy Systems Association World Congress, IFSA 1993* (Seoul, Korea, 1993), pp. 592–595

112. S. Shenoi, Fuzzy sets, information clouding and database security, in *Fuzziness in Database Management Systems, vol. 5, Studies in Fuzziness*, ed. by P. Bosc, J. Kacprzyk (Physica Publisher, Heidelberg, 1995), pp. 207–228

113. H. Stormer, N. Werro, D. Risch, An experiment on recommender systems for SME online shops. in *Proceedings of the 7th International Working For E-Business Conference, We-B 2006* (Melbourne, Australia, 2006), pp. 150–157

114. S. Stormer, N. Werro, D. Risch, recommending products by the mean of a fuzzy classification, in *Proceedings of the European Conference on Collaborative Electronic Commerce Technology and Research, CollECTeR 2006* (Basel, Switzerland, 2006), pp. 29–37

115. V. Tahani, A conceptual framework for fuzzy query processing: a step toward very intelligent database systems. Inf. Process. Manage. **13**, 289–303 (1977)

116. Y. Takahashi, A fuzzy query language for relational databases, in *Fuzziness in Database Management Systems, vol. 5, Studies in Fuzziness*, ed. by P. Bosc, J. Kacprzyk (Physica Publisher, Heidelberg, 1995), pp. 365–384

117. R. Terlutter, Verhaltenswissenschaftliche Beiträge zur Gestaltung von Kundenbeziehungen, in *Grundlagen des CRM—Konzepte und Gestaltung*, 2nd edn. ed. by H. Hippner, K.D. Wilde (Gabler Publisher, Wiesbaden, 2006), pp. 269–290
118. U. Thole, H.-J. Zimmermann, P. Zysno, On the suitability of minimum and product operators for the intersection of fuzzy sets. Fuzzy Sets Syst. **2**, 167–180 (1979)
119. E. Turban, D. King, J.K. Lee, D. Viehland, *Electronic Commerce—A Managerial Perspective*, 3rd edn. (Prentice Hall, London, 2004)
120. J. Ullman, *Principles of Database Systems* (Computer Science Press, Maryland, 1982)
121. J. Ullman, *Principles of Database Systems and Knowledge-Base Systems* (Computer Science Press, New York, 1988)
122. B. Werners, Aggregation models in mathematical programming, in *Mathematical Models for Decision Support*, ed. by G. Mitra (Springer, Berlin, 1988), pp. 295–305
123. N. Werro, A. Meier, C. Mezger, G. Schindler, Concept and implementation of a fuzzy classification query language. in *Proceedings of the International Conference on Data Mining, DMIN 2005, World Congress in Applied Computing* (Las Vegas, USA, 2005), pp. 208–214
124. N. Werro, H. Stormer, D. Frauchiger, A. Meier, eSarine—a struts-based webshop for small and medium-sized enterprises. in *Proceedings of the Information Systems in E-Business and E-Government Conference, EMISA 2004* (Luxembourg, 2004), pp. 13–24
125. N. Werro, H. Stormer, A. Meier, Personalized discount—a fuzzy logic approach. in *Proceedings of the 5th International Federation for Information Processing Conference on eBusiness, eCommerce and eGovernment, I3E 2005* (Poznan, Poland, 2005), pp. 375–387
126. N. Werro, H. Stormer, A. Meier, A hierarchical fuzzy classification of online customers. in *Proceedings of the IEEE International Conference on e-Business Engineering, ICEBE 2006* (Shanghai, China, 2006), pp. 256–263
127. N. Werro, H. Stormer, M. Savini, eSarine—Le Magasin Electronique pour PME. in *Proceedings of the Congràs International Francophone en Entrepreneuriat et PME, CIFEPME 2006* (Fribourg, Switzerland, 2006)
128. E.A. Wong, A statistical approach to incomplete information in database systems. ACM Trans. Datab. Syst. **7**(3), 470–488 (1982)
129. R.R. Yager, On a general class of fuzzy connectives. Fuzzy Sets Syst. **4**, 235–242 (1980)
130. S. Yasunobu, S. Miyamoto, Automatic train operation system by predictive fuzzy control. in *Industrial Applications of Fuzzy Control* ed. by M. Sugeno (North-Holland, 1985) pp 1–18
131. L.A. Zadeh, Fuzzy Sets. Inf. Control **8**, 338–353 (1965)
132. L.A. Zadeh, A fuzzy set interpretation of linguistic hedges. J. Cybern. **2**(3), 4–34 (1972)
133. L.A. Zadeh, Outline of a new approach to the analysis of complex systems and decision processes. IEEE Trans. Syst. Man Cybern. **3**, 28–44 (1973)
134. L.A. Zadeh, The concept of a linguistic variable and its application to approximate reasoning—Part I. Inf. Sci. **8**, 199–249 (1975)
135. L.A. Zadeh, The concept of a linguistic variable and its application to approximate reasoning—Part II. Inf. Sci. **8**, 301–357 (1975)
136. L.A. Zadeh, The concept of a linguistic variable and its application to approximate reasoning—Part III. Inf. Sci. **9**, 43–80 (1975)
137. L.A. Zadeh, Fuzzy sets as a basis for a theory of possibility. Fuzzy Sets Syst. **1**, 3–28 (1978)
138. L.A. Zadeh, PRUF—a meaning representation language for natural languages. Int. J. Man-Mach. Stud. **10**, 395–460 (1978)
139. H.-J. Zimmermann, *Fuzzy Sets Theory and its Applications*, 4th edn. (Kluwer Academic, Dordrecht, 2001)
140. H.-J. Zimmermann, P. Zysno, Latent connectives in human decision making. Fuzzy Sets Syst. **4**, 37–51 (1980)
141. D. Zumstein, *Customer Performance Measurement—Analysis of the Benefit of a Fuzzy Classification Approach in Customer Relationship Management*, Master's thesis (University of Fribourg, Fribourg, Switzerland, 2007)

142. D. Zumstein, N. Werro, A. Meier, Fuzzy Portfolio Analysis for Strategic Customer Relationship Management, Technical Report 07–01 (University of Fribourg, Fribourg, Switzerland, 2007)
143. A. Zvieli, P. P. Chen, Entity-relationship modeling and fuzzy databases. in *Proceedings of the Second International Conference on Data Engineering* (Los Angeles, 1986) pp. 320–327